Wearable Wireless Devices

Wearable Wireless Devices

Special Issue Editors

Qammer H. Abbasi
Hadi Heidari
Akram Alomainy

MDPI • Basel • Beijing • Wuhan • Barcelona • Belgrade • Manchester • Tokyo • Cluj • Tianjin

Special Issue Editors

Qammer H. Abbasi   Hadi Heidari   Akram Alomainy
University of Glasgow   University of Glasgow   Queen Mary University of London
UK   UK   UK

Editorial Office
MDPI
St. Alban-Anlage 66
4052 Basel, Switzerland

This is a reprint of articles from the Special Issue published online in the open access journal *Applied Sciences* (ISSN 2076-3417) (available at: https://www.mdpi.com/journal/applsci/special_issues/wearablewirelessdevices).

For citation purposes, cite each article independently as indicated on the article page online and as indicated below:

LastName, A.A.; LastName, B.B.; LastName, C.C. Article Title. *Journal Name* **Year**, *Article Number*, Page Range.

ISBN 978-3-03928-442-9 (Pbk)
ISBN 978-3-03928-443-6 (PDF)

Contents

About the Special Issue Editors

Qammer H. Abbasi (SM'16) received B.S. and M.S. degrees in electronics and telecommunication engineering from the University of Engineering and Technology (UET), Lahore, Pakistan, and a Ph.D. in electronic and electrical engineering from the Queen Mary University of London (QMUL), U.K., in 2012. In 2012, he was a Postdoctoral Research Assistant with the Antenna and Electromagnetics Group, QMUL. He is currently a Lecturer (Assistant Professor) at the James Watt School of Engineering, University of Glasgow, U.K., and Research Investigator with the Scotland 5G Center. He has contributed to over 250 leading international technical journal and peer-reviewed conference papers and eight books. He has received several recognitions for his research, which include appearance on BBC, STV, dawnnews, local and international newspapers, the cover of a MDPI journal, most downloaded articles, U.K. exceptional talent endorsement by the Royal Academy of Engineering, National Talent Pool Award by Pakistan, International Young Scientist Award by NSFC China, URSI Young Scientist Award, National Interest Waiver by USA, four best paper awards, and best representative image of an outcome by QNRF. He is an Associate Editor for *IEEE Journal of Electromagnetics, RF and Microwaves in Medicine and Biology, IEEE Sensors Journal, IEEE Open Access Antenna and Propagation*, and *IEEE Access*, and has acted as a Guest Editor for numerous Special Issues in top ranked journals.

Hadi Heidari is an Assistant Professor (Lecturer) at the James Watt School of Engineering and lead of the Microelectronics Lab (meLAB) at the University of Glasgow, U.K.. He received his Ph.D. in microelectronics from the University of Pavia (Italy) in 2015, where he worked on integrated CMOS sensory microsystems. He has authored over 140 articles in peer reviewed journals (e.g., IEEE *Solid-State Circuits Journal, Trans. Circuits and Systems I*, and IEEE *Trans. Electron Devices*) and in international conferences. He has organized several conferences, workshops, and special sessions, e.g., he is a founding chair of the U.K.-China Emerging Technology (UCET) workshop, U.K. Circuits and Systems (UKCAS) workshop, and member of the organzsing committee of SENSORS'17-'18, BioCAS'18, and PRIME'2019. He is an IEEE Senior Member, an Editor for the Elsevier Microelectronics Journal and lead Guest Editor for four journal Special Issues. He is member of the IEEE Circuits and Systems Society Board of Governors (BoG), and Member-at-Large IEEE Sensors Council. He has received several best paper awards from IEEE international conferences including ISCAS'14, PRIME'14, and ISSCC'16, and a travel scholarship from IEEE NGCAS'17. He has a grant portfolio of +£1 million, funded by major research councils and funding organizations including European Commission, U.K.'s EPSRC, the Royal Society, and Scottish Funding Council.

Akram Alomainy received a M.Eng. degree in communication engineering and a Ph.D. in electrical and electronic engineering (specialized in antennas and radio propagation) from Queen Mary University of London (QMUL), U.K., in July 2003 and July 2007, respectively. He joined the School of Electronic Engineering and Computer Science, QMUL, as a reader in antennas applied EM. His current research interests include small and compact antennas for wireless body area networks, radio propagation characterization and modelling, antenna interactions with the human body, computational electromagnetics, advanced antenna enhancement techniques for mobile and personal wireless communications, nano-scale networks and communications, THz material characterization and communication links, and advanced algorithms for smart and intelligent antennas and cognitive radio system. He has authored and co-authored four books, 6 book chapters, and more than 350 technical papers (6400+ citations and H-index of 35) in leading journals and peer-reviewed conferences. Dr. Alomainy won the Isambard Brunel Kingdom Award in 2011 for being an outstanding young science and engineering communicator. He was selected to deliver a TEDx talk about the science of electromagnetics and also participated in many public engagement initiatives and festivals. He is an elected member of the U.K. URSI (International Union of Radio Science) panel to represent the U.K. interests of the URSI Commission B (1 September, 2014 until 31 August, 2020).

Editorial

Wearable Wireless Devices

Qammer Hussain Abbasi [1,2,*], **Hadi Heidari** [1] and **Akram Alomainy** [2]

[1]   James Watt School of Engineering, University of Glasgow, Glasgow G12 8QQ, UK
[2]   School of Electronic Engineering and Computer Science, Queen Mary University of London, London E1 4NS, UK
*   Correspondence: qammer.abbasi@glasgow.ac.uk

Received: 5 June 2019; Accepted: 24 June 2019; Published: 29 June 2019

1. Introduction

With the growing interest in the use of technology in daily life, the potential of using wearable wireless devices across multiple segments, e.g., healthcare, sports, child monitoring, military, emergency, consumer electronics, etc., is rapidly increasing. It is predicted that there will be multibillion wearable sensors by 2025, with over 30% of them being new types of sensors that are just beginning to emerge. This Special Issue will be focused on wireless wearable and implantable system, flexible textile-based electronics, bio-electromagnetics, antennas and propagation, RF circuits, sensor, security of wearables and implantable system, nano-bio communication and electromagnetic sensing.

2. Contributions

The special issue consists of 10 contributions in the area of wearable wireless devices:

In the paper, 'Reservoir Computing Based Echo State Networks for Ventricular Heart Beat Classification' [1] a Reservoir Computing (RC) based Echo State Networks (ESNs) for ventricular heartbeat classification based on a single Electrocardiogram (ECG) lead has been proposed. The proposed method was especially designed for Medical Internet of Things (MIoT) devices, for instance wearable wireless devices for ECG monitoring or ventricular heart beat detection systems and so on. The performance of the proposed model was evaluated using the MIT-BIH-AR dataset and it achieved excellent results.

In the paper, 'A Real Time and Lossless Encoding Scheme for Patch Electrocardiogram Monitors' [2] an ECG encoding scheme for joint lossless data compression and heartbeat detection to minimize the circuit footprint size and power consumption of a patch electrocardiogram (ECG) monitors (PEMs) has been proposed. The proposed encoding scheme supports two operation modes: fixed-block mode and dynamic-block mode. Dynamic-block mode provides heartbeat detection accuracy at a rate higher than 98%. Fixed-block mode was also implemented on the field-programmable gate array, and could be used as a chip for using analog-to-digital convertor-ready signals as an operation clock.

In the paper, 'Foot-Mounted Inertial Measurement Units-Based Device for Ankle Rehabilitation' [3], authors presented an inertial measurement units (IMU)-based physical interface for measuring the foot attitude, and a graphical user interface that acts as a visual guide for patient rehabilitation. According to the results, more consistent rehabilitation could be achieved by providing feedback on foot angular position during therapy procedures.

In the paper, 'Chronic Obstructive Pulmonary Disease Warning in the Approximate Ward Environment' [4] the usage of modern 5G C-Band sensing for health care monitoring has been proposed. The focus of this research was to monitor the respiratory symptoms for COPD (Chronic Obstructive Pulmonary Disease). The 5G sensing technique enhances the sensing performance for the health care sector by monitoring the amplitude information for different respiratory activities of a patient using the above-mentioned devices.

In the paper, 'Movement Noise Cancellation in Second Derivative of Photoplethysmography Signals with Wavelet Transform and Diversity Combining' [5] authors proposed an algorithm to remove movement noise from second derivative of photoplethysmography (SDPPG) signals. Experiment results show that the proposed algorithm outperforms the previous filter-based algorithm, and that movement noise with 30% time duration can be reduced by up to 70.89%.

In the paper, 'An Anonymous Mutual Authenticated Key Agreement Scheme for Wearable Sensors in Wireless Body Area Networks', [6] authors proposed that Li et al. lightweight protocol for wearable sensors in wireless body area networks is still vulnerable to three types of attacks i.e., the offline identity guessing attack, the sensor node impersonation attack and the hub node spoofing attack. Authors present a secure scheme that addresses these problems, and retains similar efficiency in wireless sensors nodes and mobile phones.

In the paper, 'Respiration Symptoms Monitoring in Body Area Networks' [7] authors presented a framework that monitors particular symptoms such as respiratory conditions (abnormal breathing pattern) experienced by hyperthyreosis, sleep apnea, and sudden infant death syndrome (SIDS) patients. The rhythmic patterns extracted using S-Band sensing present the periodic and non-periodic waveforms that correspond to normal and abnormal respiratory conditions, respectively.

In the paper, 'Internet of Things for Sensing: A Case Study in the Healthcare System' [8], authors did a pilot study, to look at narcolepsy, a disorder in which individuals lose the ability to regulate their sleep-wake cycle. Using S-band sensing Classification and validation of various human activities such as walking, sitting on a chair, push-ups, and narcolepsy sleep episodes are done using support vector machine, K-nearest neighbor, and random forest algorithms. The measurement and evaluation were carried out several times with classification values of accuracy, precision, recall, specificity, Kappa, and F-measure of more than 90% that were achieved when delineating sleep attacks.

In the paper, 'Emergency-Prioritized Asymmetric Protocol for Improving QoS of Energy-Constraint Wearable Device in Wireless Body Area Networks' [9], authors proposed a new MAC protocol to satisfy the higher energy efficiency of nodes than coordinator by designing the asymmetrically energy-balanced model between nodes and coordinator. The proposed scheme loads the unavoidable energy consumption into the coordinator instead of the nodes to extend their lifetime. Additionally, the scheme also provides prioritization for the emergency data transmission with differentiated Quality of Service (QoS). For the evaluations, IEEE 802.15.6 was used for comparison.

In the paper, 'Virtual Reality-Wireless Local Area Network: Wireless Connection-Oriented Virtual Reality Architecture for Next-Generation Virtual Reality Devices' [10], authors carefully examine the feasibility of wireless VR over WLANs, and proposed an efficient wireless multiuser VR communication architecture, as well as a communication scheme for VR. Extensive simulations have been performed to corroborate the outstanding performance of the proposed scheme.

The editors hope that this special issue will benefit the scientific community and contribute to the knowledge base.

Acknowledgments: The editors would take this opportunity to applaud the contribution of the authors to this special issue. Efforts of the reviewers to enhance the quality of the manuscripts is also much appreciated.

Conflicts of Interest: The authors declare no conflict of interest.

References

1. Mastoi, Q.; Wah, T.; Gopal Raj, R. Reservoir Computing Based Echo State Networks for Ventricular Heart Beat Classification. *Appl. Sci.* **2019**, *9*, 702. [CrossRef]
2. Fang, H.; Lu, C. A Real Time and Lossless Encoding Scheme for Patch Electrocardiogram Monitors. *Appl. Sci.* **2018**, *8*, 2379. [CrossRef]
3. Gómez-Espinosa, A.; Espinosa-Castillo, N.; Valdés-Aguirre, B. Foot-Mounted Inertial Measurement Units-Based Device for Ankle Rehabilitation. *Appl. Sci.* **2018**, *8*, 2032. [CrossRef]
4. Zhang, Q.; Haider, D.; Wang, W.; Shah, S.; Yang, X.; Abbasi, Q. Chronic Obstructive Pulmonary Disease Warning in the Approximate Ward Environment. *Appl. Sci.* **2018**, *8*, 1915. [CrossRef]

5. Ban, D.; Shahid, S.; Kwon, S. Movement Noise Cancellation in Second Derivative of Photoplethysmography Signals with Wavelet Transform and Diversity Combining. *Appl. Sci.* **2018**, *8*, 1531. [CrossRef]

6. Chen, C.; Xiang, B.; Wu, T.; Wang, K. An Anonymous Mutual Authenticated Key Agreement Scheme for Wearable Sensors in Wireless Body Area Networks. *Appl. Sci.* **2018**, *8*, 1074. [CrossRef]

7. Liu, L.; Shah, S.; Zhao, G.; Yang, X. Respiration Symptoms Monitoring in Body Area Networks. *Appl. Sci.* **2018**, *8*, 568. [CrossRef]

8. Shah, S.; Ren, A.; Fan, D.; Zhang, Z.; Zhao, N.; Yang, X.; Luo, M.; Wang, W.; Hu, F.; Rehman, M.; et al. Internet of Things for Sensing: A Case Study in the Healthcare System. *Appl. Sci.* **2018**, *8*, 508. [CrossRef]

9. Lee, J.; Kim, S. Emergency-Prioritized Asymmetric Protocol for Improving QoS of Energy-Constraint Wearable Device in Wireless Body Area Networks. *Appl. Sci.* **2018**, *8*, 92. [CrossRef]

10. Ahn, J.; Kim, Y.; Kim, R. Virtual Reality-Wireless Local Area Network: Wireless Connection-Oriented Virtual Reality Architecture for Next-Generation Virtual Reality Devices. *Appl. Sci.* **2018**, *8*, 43. [CrossRef]

 *applied sciences*

Article

Reservoir Computing Based Echo State Networks for Ventricular Heart Beat Classification

Qurat-ul-ain Mastoi, Teh Ying Wah * and Ram Gopal Raj

Faculty of Computer Science and Information Technology University of Malaya, Kuala Lumpur 50603, Malaysia; quratulain.mastoi@siswa.um.edu.my (Q.-u.-a.M.); ramdr@um.edu.my (R.G.R.)
* Correspondence: tehyw@um.edu.my

Received: 15 January 2019; Accepted: 1 February 2019; Published: 18 February 2019

Abstract: The abnormal conduction of cardiac activity in the lower chamber of the heart (ventricular) can cause cardiac diseases and sometimes leads to sudden death. In this paper, the author proposed the Reservoir Computing (RC) based Echo State Networks (ESNs) for ventricular heartbeat classification based on a single Electrocardiogram (ECG) lead. The Association for the Advancement of Medical Instrumentation (AAMI) standards were used to preprocesses the standardized diagnostic tool (ECG signals) based on the interpatient scheme. Despite the extensive efforts and notable experiments that have been done on machine learning techniques for heartbeat classification, ESNs are yet to be considered for heartbeat classification as a is fast, scalable, and reliable approach for real-time scenarios. Our proposed method was especially designed for Medical Internet of Things (MIoT) devices, for instance wearable wireless devices for ECG monitoring or ventricular heart beat detection systems and so on. The experiments were conducted on two public datasets, namely AHA and MIT-BIH-SVDM. The performance of the proposed model was evaluated using the MIT-BIH-AR dataset and it achieved remarkable results. The positive predictive value and sensitivity are 98.98% and 98.98%, respectively for the modified lead II (MLII) and 98.96% and 97.95 for the V1 lead, respectively. However, the experimental results of the state-of-the-art approaches, namely the patient-adaptable method, improved generalization, and the multiview learning approach obtained 92.8%, 87.0%, and 98.0% positive predictive values, respectively. These obtained results of the existing studies exemplify that the performance of this method achieved higher accuracy. We believe that the improved classification accuracy opens up the possibility for implementation of this methodology in Medical Internet of Things (MIoT) devices in order to bring improvements in e-health systems.

Keywords: reservoir computing; echo state networks; bio signals, bio sensors; heart beat classification; medical internet of things; medical wearable wireless devices

---

1. Introduction

According to the American Heart Association [1] and the World Health Organization [2], the mortality ratio due to heart diseases is growing rapidly in the world. It is estimated that around 17.7 million people were affected due to cardiovascular disease (CVD) in 2015. The documentation report from China in 2011 [3] disclosed the information of CVD patients, which showed the statistics that around 230 million people were suffering from CVD and that 3 million death cases were due to the CVD. Electrocardiogram (ECG) signals are a major tool which have been widely used for the analysis of heart disorders in many applications [4–6]. The collection of these ECG signals are based on contemporary devices which are being used to communicate with the body surface using sensors or nodes [7]. The ECG signals refers to the physical activity of the heart; either the heart has disease or an abnormal rhythm. Irregular rhythm or abnormal rhythm is also known as arrhythmia, which is the leading source of any heart disease. It is not stated that all arrhythmias are dangerous, however it requires therapy to avoid the severe heart diseases [8]. In recent years, many computer-assisted techniques have been proposed to classify heartbeats for arrhythmia prediction [8–10]. The two main aspects were considered in previous studies (1) feature extraction techniques were used to extract the morphological changes from the ECG signals and, (2) learning algorithms for the classification of heart beats. Based on the feature extraction aspect, the studies used time domain analysis [8,11,12], frequency domain analysis [13–15], and feature analysis based on wavelet transforms [16–19]. The most common selection of learning algorithms includes Support Vector Machines [11,12,14,20], K-nearest Neighbors [21,22], Neural Networks [19,23–26], decision trees [15,27], and deep learning [28–30].

With the increasing trend of contemporary ECG monitoring devices, the demands of efficient computer-assisted diagnostic methods are also increasing. These types of devices demand fully automated methods for real-time classification which requires less computational cost and can easily be adaptable for hardware. Nowadays, the concept of the Medical Internet of Things is improving the e-health systems. All of these systems require less computational cost and time-saving algorithms for classification. However, despite the extensive efforts that have been made in the previous techniques for heartbeat classification with the real-time scenarios, they often deal with a limited number of leads which are often contaminated and some of the studies are lead specific which do not perform efficiently in terms of cost and time [12].

Furthermore, inter-patient variability is the major problem in ECG signals because heart beats with the same class behave differently in patients. In the literature, most of the studies were majorly focused on simple heartbeat classification based on the single lead ECG signal. Therefore, we observed that the analysis of ECG signals are not sufficient to classify the proper heartbeats; to bridge this gap, the author proposed the patient adaptable approach template matching [31] which allows a learning algorithm to classify the heartbeat lead-specific and this method also enhances the efficiency of the algorithm to classify heartbeats from non-standard leads.

In this work, this paper proposed the Reservoir computing based echo state-network for ventricular heartbeat classification. The design of the Reservoir computing is based on a large recurrent neural network (RNN), and the main advantage of this proposed ring network is to handle real-world problems efficiently with high processing speed. This study used ESN (Echo State Networks) with non-random cyclic technique to classify the ventricular heartbeats from the physiological signals. The main advantage of this technique is that the output connections are fitted by using the simple regression technique [32] and only the network outputs are being trained [33]. The other main reason to apply Reservoir Computing Based ESNs is that it is easy to handle large real-time data and it is also adjustable to implement in medical devices, for instance, Medical IoT gadgets.

2. Materials and Methods

The workflow of our proposed methodology is illustrated in Figure 1. The signal pre-processing stage comprises of several steps, which include denoising of the ECG signals, peak detection, heartbeat segmentation, and morphological, temporal feature extraction. Finally, the model is evaluated based on the input materials (processed feature sets) by training and testing the classifier.

2.1. ECG Data Sources

The experiment was conducted based on two different databases named the American Heart Association Ventricular Arrhythmia ECG database (AHA) [34] and Massachusetts Institute of Technology-Beth Israel Hospital Supraventricular Arrhythmia database (MIT-BIH SVDB) [35]. These databases are used to train the study's proposed classifier. On the other side, the paper used MIT-BIH arrhythmia to evaluate the performance of our proposed classifier. The list of all acronyms which are used in this study is present in Table 1 and the overall description of these above-mentioned datasets is defined in Table 2. According to the American National Standards Institute/Association for the Advancement of Medical Instrumentation (ANSI/AAMI) standards, the ECG signal recordings have annotated beats in five groups [36]: Ventricular ectopic beat (VEB), Supraventricular ectopic beat (SVEB), Fusion beat (F), Unclassified and paced beat (Q), and Non-ectopic beat (N).

Table 1. List of acronyms used in this study.

Acronyms	
MIT-BIH-AR	Massachusetts Institute of Technology-Beth Israel arrhythmia
MIT-BIH-SVDM	Massachusetts Institute of Technology-Beth Israel Hospital Supraventricular Arrhythmia database
AHA	American Heart Association
ECG	Electrocardiogram
RC	Reservoir computing
MIoT	Medical Internet of Things
ANDSI/AAMI	American Nation Standards Institute/Association for the Advancement of Medical Instrumentation
SVB	Supraventricular Beat
VB	Ventricular Beat

However, paced beat recordings were omitted from the datasets. These annotated signal-based ECG recordings have a 30 min long duration, whereas the MIT-BIH-SVDM dataset has two types of lead recordings: the modified lead II (MLII) and V1; a few patients also have V5 and V2 instead of V1. The AHA dataset has no information about the lead recordings. Hence, this study's methodology is based on the classification of a ventricular heartbeat. Therefore, in this paper we consider two classes of AAM/ANSI standards for our binary class models to determine the normal and abnormal beats: Supraventricular beats as normal morphology and has a label 0, whereas the ventricular beats as abnormal morphology and has a label 1.

Table 2. The Description of Data sets used in this study.

Dataset	ECG Rec	Patients	Leads	SVB	VB
AHA	155	-	2	317,612	32,403
MIT-BIH-SVDB	78	-	2	174,317	9953
MIT-BIH-AR	48	47	2	92,754	7803

2.2. ECG Signal Preprocessing

In this study, ECG signals-based recordings are used, and these are discussed in the previous Section 2.1. The signals are collected based on the common sampling frequency rate of 360 Hz. In order to create a new feature set for input of the classifier, by using the ECG signal dataset, the following steps were conducted and summarized in Figure 1.

2.2.1. ECG Signal Denoising

The contaminated signals contain numerous kinds of noises and artifacts, for instance, baseline wanderings and power line interference. There are several reasons for contamination; sometimes it occurs due to respiration or it also occurs when the patient moves while recording the physiological signals. There are two noises that are frequently highlighted in ECG signals baseline wandering and Power line interference [37] which are caused by respiration and the variation of 50% amplitude in peak to peak due to the 50 Hz, respectively. These noises and artifacts can create a problem in the extraction of information of interest (hidden features) from the raw ECG signal. Furthermore, a corrupt ECG signal can lead to the wrong diagnosis and it also has a major effect on the performance of algorithms during classification [37–39].

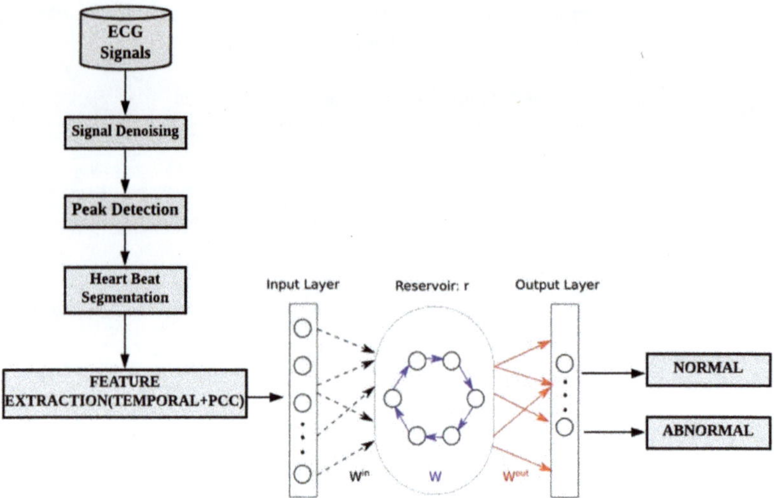

Figure 1. Schematic representation of ventricular heartbeat classifier.

In the literature, there are many studies that are dedicated to innovating the algorithm for the filtering of noisy physiological signals for the highest performance of proposed models, for instance [39,40], this study applies the same filtering technique as used in [8,41], and these studies also addressed the similar aim of classification. The author believes that this is the major decision that helps to evaluate the performance of the current study with the existing methods. Therefore, to bridge this gap, we used a median filter on the ECG signal (x) which had *n*th length. The sliding window is created for the preprocessing of signals, so the length of the ECG signal n will be represented as the length of the sliding window. For every step, the median value will be calculated by using this equation:

$$\begin{cases} x\left(i - \frac{n-1}{2}\right) : x\left(i + \frac{n-1}{2}\right) \; \textit{if the value of n is odd} \\ x\left(i - \frac{n}{2}\right), x\left(i - \left(\frac{n}{2}\right) + 1, \ldots, x\left(i + \left(\frac{n}{2}\right) - 1\right) \; \textit{if the value of n is even} \end{cases} \tag{1}$$

This study used two median filters, one is for length $n = 600$ ms and the second one is for $n = 200$ ms. This study used the medfilt1 function of the Matlab [42] which implemented the one-dimensional filtering on the ECG signal. The main benefit of applying the medfilt1 function was that it eliminates the unwanted distortion from the signal.

After the application of median filters for the baseline wandering outlier's removal, this study applied 12th order finite impulse response (low pass filter) with the given cut-off frequency $k = 35$ Hz for removing the outliers which are related to power line interference by using the fir1 Matlab function [43]. The clear picture based on the results of these filters is reported in Figure 2a,b.

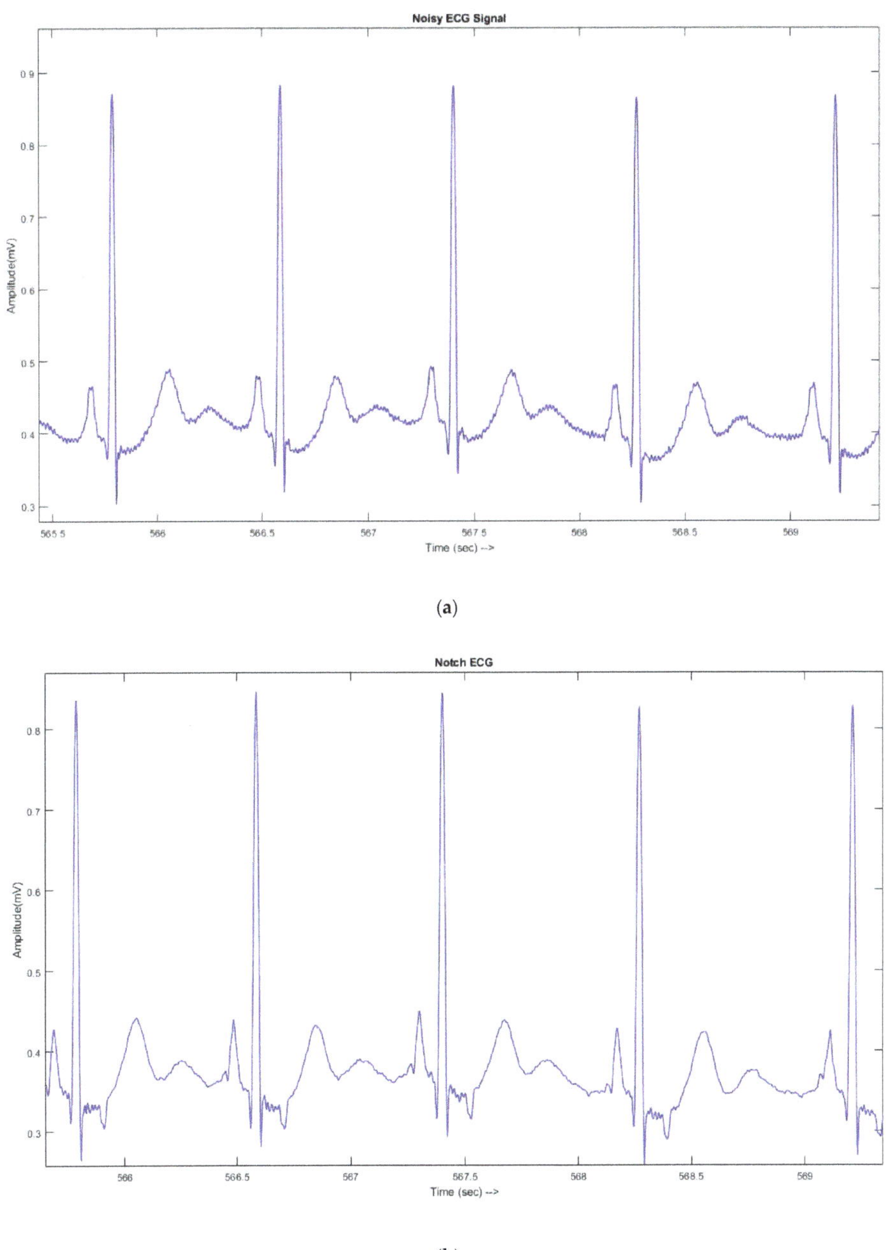

(a)

(b)

Figure 2. (**a**) Representation of Noisy ECG signal; (**b**) Representation of Filtered ECG signal.

2.2.2. High Frequency component (Peak) Detection

This study carried out the second step peak detection which was very useful for further steps.

The High-frequency components are detected by applying the modified Pan Tompkins algorithm [44]. The algorithm consists of the following steps:

In the first step, this study used a block of the differentiation equation to obtain the high slope values; secondly, the output values of the signals were squared to extract the R-peaks; in the last step, this study applied the summation on all values of R-wave slope. The values related to the R-wave, for instance, R-location and amplitudes of R-wave were stored in the $\mathbf{R}^{nXm}$ matrix. Figure 3 represents the example of detecting the R-peaks based on the Pan Tompkins algorithm. The reason for applying this modified Pan Tompkins algorithm is that it can easily adapt the variation of signal changes and extract the high-frequency component in an efficient manner.

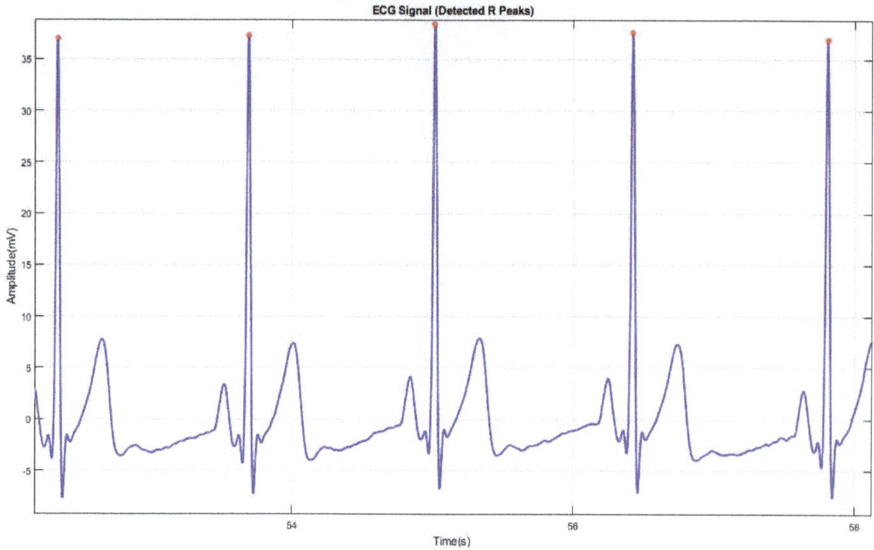

Figure 3. Representation of detected R-peaks in the ECG signal.

2.2.3. Heart Beat Segmentation

The aim of this subsection is to segment the heartbeat; after the detection of peaks, this study used the $\mathbf{R}^{nxm}$ matrix for selecting the R-location as a reference point. After determining the position of R-peak, 200 samples or around 0.52 s points were separated in the window, which is considered a single heartbeat. The segmented area based on the QRS complex is described in Figure 4. The start point and end point of the highlighted beat represented as Q wave and S wave respectively, whereas the peak area is defined as R wave. The combination of Q, R and S wave represents ventricular depolarization of the heart.

However, for the extraction of P and T waves, this study divided the 200 samples in two part (the division of points was based on 75 and 125 points which were taken from the left side and right side of the R-peak, respectively). To avoid the misleading of extraction, all samples were collected from 0.2 s to 0.32 s. In this way, using the temporal location and samples, the study extracted the Q, S, P, and T waves from each heartbeat.

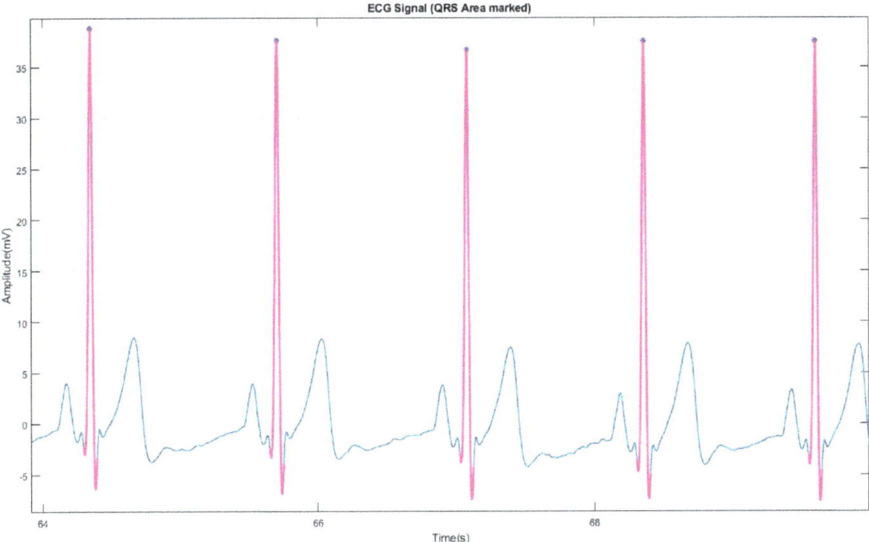

Figure 4. Representation of QRS segmented area in the ECG signal.

2.3. Temporal Feature Extraction

The aim of this study is to design a reliable and optimal classifier for ventricular heartbeat classification which can be useful for Medical IoT based on a single ECG lead. The other main focus of this study is to extract those features which should not contain more computational cost and should be easy to implement in the real time environment. Therefore, this study does not consider those signals for the feature extraction phase which has contamination in their parts.

The characterization of normal heart beats (N + SVB) and abnormal heartbeats (A + VB) are well known and are also suggested by the medical specialists in the literature [45,46]. (N + SVB) were distinguished by regular RR intervals, the presence of P-wave, and has a narrow QRS complex, whereas (A + VB) has shorter RR interval, the p-wave is not present, and it also has a wider QRS complex.

Moreover, in this part of feature extraction, this study follows two types of methods; one is based on temporal feature extraction which is less expensive and easy to implement, however these features are lead independent yet useful for real-time implementation, therefore to overcome this inter-patient variability issue, this study used a second method by the name template matching. It is a patient adaptable and simple approach which measures the similarity between the input signal beat and template beat [47].

In the first method, the author computed the relevant attributes for the study's classifier; the six temporal features were computed from each segmented beat.

1. **The Previous R-R interval** is defined as the time duration between the current beat and previous beat.
2. **The Subsequent R-R Interval** is explained as the time duration between the current beat and subsequent beat.
3. **The Standard deviation of Successive Difference (SDSD)** is defined as the standard feature of physiological signals for arrhythmia classification. It is the difference between 10 consequent RR intervals [48].
4. **The Average of RR-Interval** $\left|\overline{Beat}\right|$ is defined as the average ratio of 10 consecutive RR intervals.

5. **The Average Derivate of RR interval** $\left|\overline{Beat}'\right|$ is defined as the average ratio of the derivate of the RR interval, where the derivation of the segmented beat is calculated by using this central difference Equation (2).

$$\left(f'(z) \approx \frac{f(z+h) - f(z-h)}{2h} \right) \tag{2}$$

Based on clinical specialist advice, ventricular abnormal beat has a wider area of QRS complex and it also tends to rise and fall more than a normal beat. Therefore, the derivative area should have the lowest value.

In contrast, in the second method, template matching was used to avoid the lead independency and inter-patient variability issues. The approach was quite simple and patient adaptable; the study used this approach to compare the ECG signal heart beats with the same class specific template in order to define the Pearson's correlation coefficients (PCC) between the input ECG signal and template beat. However, template beat was calculated by taking an average ratio of at least 50% beats in which 25% belongs to the normal beat and 25% belongs to the abnormal beats. After calculating the specific template beats for normal and abnormal, this method compared the current heartbeat with the template beat example which is shown in Figure 5.

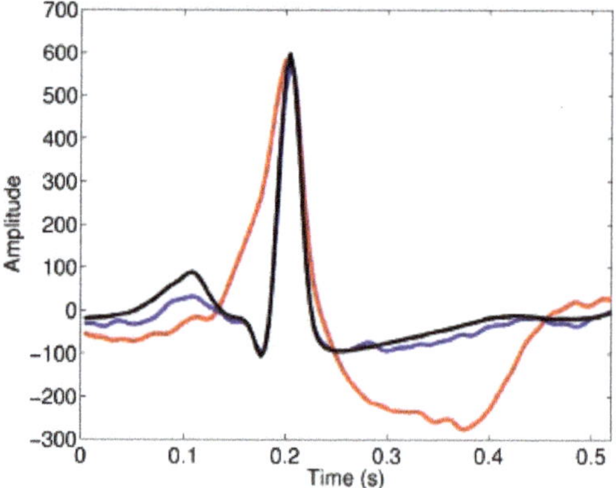

Figure 5. Template matching of segmented heartbeat where red is the original abnormal beat and blue and black is the original normal beat and template beat, respectively.

To this end, for the comparison of heartbeats in real time scenarios, we updated and calculated the template based on a real-time adaptation of the system. Assuming that doctors are using 10 s ECG strips which have around 15 heartbeats in every 10 s ECG strips, the new template is estimated and updated and also replaces the old beats from the template. To evaluate the similarity measurement between the template beat and ECG signal heartbeat (PCC), features are extracted for our classifier as follows:

6. **PCC** (*Template beat, heart beat*) is defined as the Pearson's correlation coefficients between the template beat and ECG signal heartbeat.
7. **PCC (** *Template beat', heart beat'*) is defined as the Pearson's correlation coefficients between the derivatives value of the template beat and ECG signal segmented beat.
8. **PCC (***Template beat^2, heart beat^2*) is defined as the Pearson's correlation coefficients between the squares ratio of the template beat and ECG signal segmented beat.
9. **PCC** $\left((Template\ beat')^2, (heart\ beat')^2 \right)$ is defined as the Pearson's correlation coefficients between the squares of the derivatives of the template beat and ECG signal segmented beat.
10. **The Average of Template beat** $\left|\overline{Template\ beat}\right|$ is defined as the average of the template beat.

11. **The Average Derivate of Template beat** $\left|\overline{Template\ beat}'\right|$ is defined as the average derivatives of the template beat.

To this end, the total 11 features are selected for the study proposed classifier in which five were based on temporal features of ECG morphology and the rest of the features were based on (PCC) by using the template matching approach. All these features are stored in the matrix denoted as $F(n)$.

2.4. The Configuration of the Echo State Network Based Reservoir Computing for Classification

The main idea of reservoir computing (RC) is based on artificial neural network (ANN) because the nature of functional elements in RC is based on the input weights, biases, and connection weights among the neurons that are similar to ANN. The RC is further divided into two common methods (1) Echo State Networks (ESNs) [32] and (2) liquid state machine (LSMs) [49]. ESNs tend to use sparsely connected sigmoid nodes, whereas LSMs are a network which is made up by spiking neurons in the reservoir model [50]. In this study, the author considered the main type of RC echo state networks (ESNs) to construct an efficient model; for implementation in real-time scenarios, see Figure 6. In our methodology, the author used a cycle-based ESN architecture where the neurons of the reservoirs were connected in a ring form, which is also called non-random links between neurons (see Figure 7).

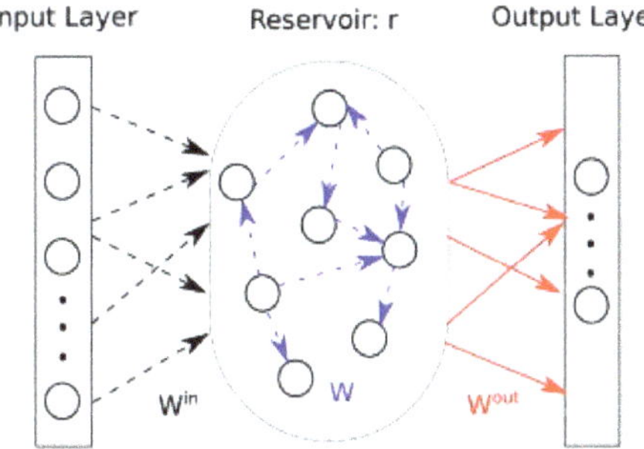

Figure 6. Representation of standard ESN (Echo State Networks) architecture.

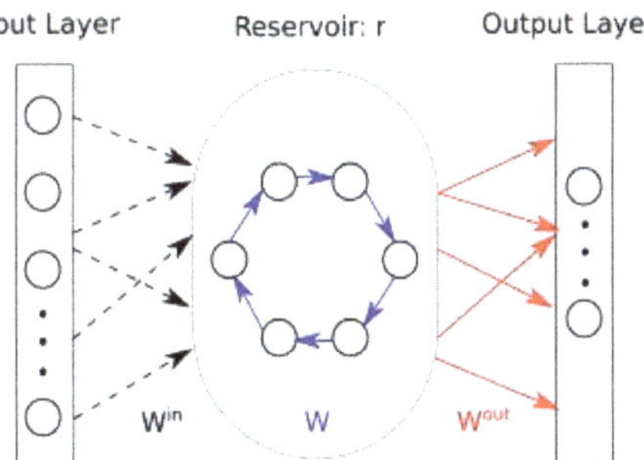

Figure 7. Representation of standard proposed cyclic or ring ESN architecture.

The reason behind using the non-random topology was that an ESNs model could easily be implemented in hardware and it also allows low power consumption with efficient processing speed [51]. These advantages encouraged the author to use the ESNs model for ventricular heart beat classification. The activation vector of the echo state network is defined by Equation (3).

$$s(n) = g(\vartheta W^{in} F(n) + \alpha W s(n-1)) \tag{3}$$

where $S(n)\epsilon\mathbb{R}^{N_d}$, is the activation vector or state in which N_x represents the number of neurons in the connection weights and $W^{in}\epsilon\mathbb{R}^{N_x X\ N_d}$ is a random matrix which was typically drawn uniformly from $[-1, 1]$ in which N_d represents the dimension of the input vector. In the end, the connection weight matrix was computed based on the connection between neurons in the network and it is represented as $W\epsilon\mathbb{R}^{N_x X\ N_x}$, whereas ϑ and α are the input parameters of connection scaling, $F(n)$ represent the input feature vector of heart beats which has dimensionality = 11. The activation function of ESN is also called classic sigmoid function which is shifted to become symmetric near to 0. In general, the ESN state initialized with null state. The linear combination of the N_x activation $s(n)$ or response of the ESN model was calculated according to:

$$y(n) = g\big(W^{out}S(n)\big) \tag{4}$$

where $W^{out}\epsilon\ \mathbb{R}^{N_{out} X\ N_x}$ represent the weight matrix of the connections between ESN neurons and output nodes, N_{out} represents the number of readouts. We extend our ESN model to know the bias weight and feedback between the $y(n)$ and the reservoirs is defined as:

$$S(n) = g\Big(W^{in}\ F(n) + W_{s(n-1)} + W^{OR}_{y(n-1)} + W^{bias}\Big) \tag{5}$$

2.4.1. Learning Phase of The ESN Model

In the learning phase of this study, the author split our dataset based on a 10-fold cross-validation technique. The substantial amount of data around 80% was used to train the classifier, whereas 20% of data was used for testing. In this training phase, the ESN model tries to find the weight matrix which minimizes the error rate between the output and target values. The representation of the training input and output sequence is defined as:

$$C = \Big(W^{in}(1), W^{out}\ (1)\Big),\ldots\ldots\ldots,\Big(W^{in}(n)_{max}), W^{out}\ (n_{max})\Big) \tag{6}$$

Hence, the output matrix $W^{out}\epsilon\ \mathbb{R}^{N_{out}\ X\ N_x}$ and the reservoir states in the training phase are represented as $A\ \epsilon\mathbb{R}^{N_x\ X\ T}$. The corresponding output weight matrix is defined as $B\ \epsilon\mathbb{R}^{N_x\ X\ T}$, where T denotes the time of training phase. The training is defined as follows:

$$W^{out}\ A = B \tag{7}$$

It can be rewritten as

$$W^{out}\ AA^T = BA^T \tag{8}$$

The solution of this equation is computed as in inverse matrix in order to find out that W^{out} is defined as:

$$W^{out} = \big(BA^T\big)\big(AA^T\big)^{-1} \tag{9}$$

We used Moore Penrose pseudoinverse [52] for numerical stability which is defined as:

$$W^{out} = BA^T\big(AA^T\big)^{+} \tag{10}$$

To mitigate the overfitting of the model, we used Ridge Regression [53] which minimized the amplitude of W^{out} which is according to:

$$W^{out} = BA^T\big(AA^T + \lambda I\big)^{-1} \tag{11}$$

where λ is the regularization factor of each reservoir which has no absolute meaning and I represents the identity matrix $\big(I \in \mathbb{R}^{N_x\ X\ N_x}\big)$.

The ratio of normal and abnormal values is estimated here in this point based on scaling parameters, the maximum threshold value of the study set is 0.4, and by using this value, the study scales our parameters to get the accurate prediction. In the end of the training phase, the regression technique helps to minimize the quadratic error between the output values and desired output values. The performance of our proposed classifier achieved outer perform results which are discussed in the following section.

3. Results and Discussion

3.1. Performance Evaluation Parameters

To evaluate the performance of our ESN model, the study used three performance parameters which are recommended by AAMI standards for evaluating the performance of learning algorithm, which includes sensitivity (Se), classification accuracy (Acc), and error rate (Er)

$$S_e = \frac{TP}{TP + FN} * 100 \tag{12}$$

$$Acc = \frac{TP + TN}{TP + TN + FP + FN} * 100 \tag{13}$$

$$E_r = \frac{FN + FP}{N} \tag{14}$$

where **True negative (TN)** is defined as the number of normal records, which is correctly classified as normal, **True positive (TP)** is defined as the number of abnormal records, which is correctly classified as abnormal, **False positive (FP)** is defined as the number of normal records that are classified as an abnormal record of the dataset, and **False Negative (FN)** is defined as the number of abnormal records that are classified as normal.

3.2. Evaluation

In this section, the performance of the proposed classifier is evaluated in terms of accuracy, sensitivity, and error rate. This study can also indicate that this evaluation part is the final classifier performance as the study already trained the model before. Table 3 shows that ECG 1 and ECG 2 from (MIT-BIH SVDB) and AHA recordings obtained remarkable performance. The ECG 1 lead achieved 98% accuracy, whereas ECG 1 and AHA recordings obtained 97% and 96% accuracy, respectively. ECG 1 achieved the highest result because this lead is modified which is good enough to distinguish the beats very efficiently. The detailed performance of our proposed model is described in Figures 8–11 where the confusion matrix shows the total number of classified beats and the graph represents the performance statistics based on percentage in terms of accuracy and sensitivity.

Table 3. Performance Analysis of ESN model.

Leads	Acc (%)	Se (%)	Er
ECG 1	98	98.98	0.02
ECG 2	97	98.97	0.03
AHA	96	97.95	0.04

Where acc is accuracy, Se is sensitivity, Sp is specificity, and Er is the error rate.

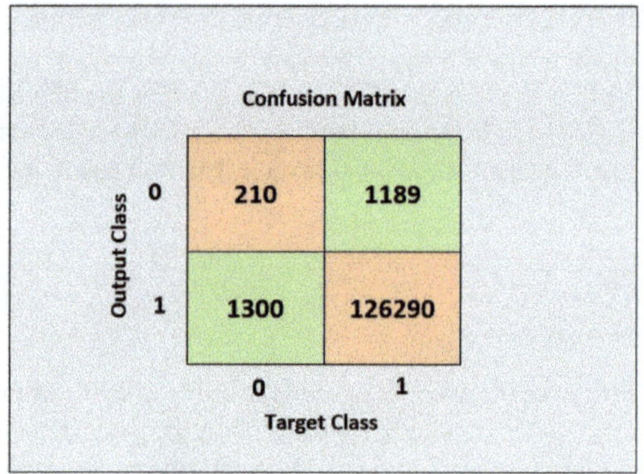

Figure 8. Confusion matrix of the ECG1 classified beats.

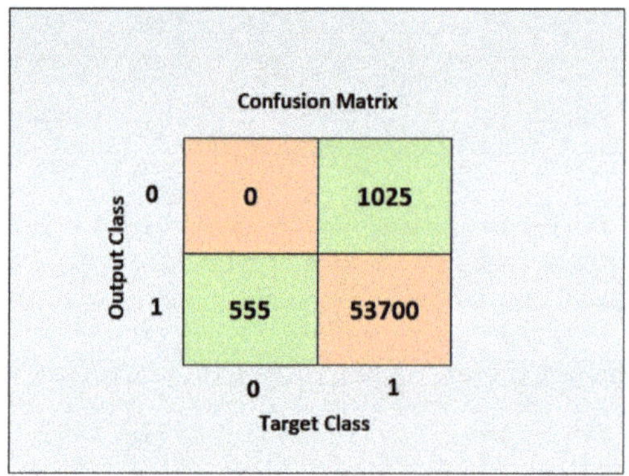

Figure 9. Confusion matrix of the ECG2 classified beats.

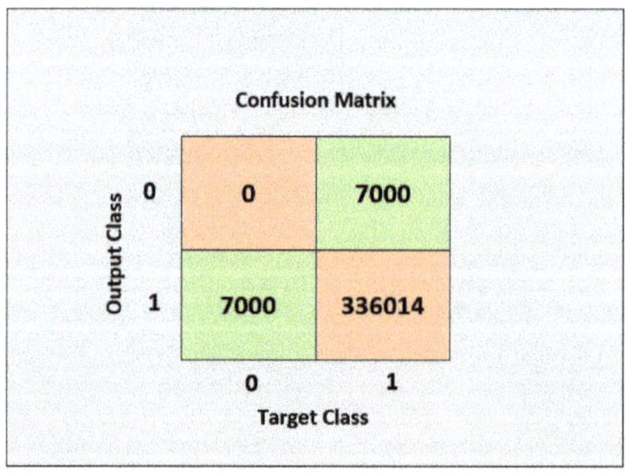

Figure 10. Confusion matrix of the AHA classified beats.

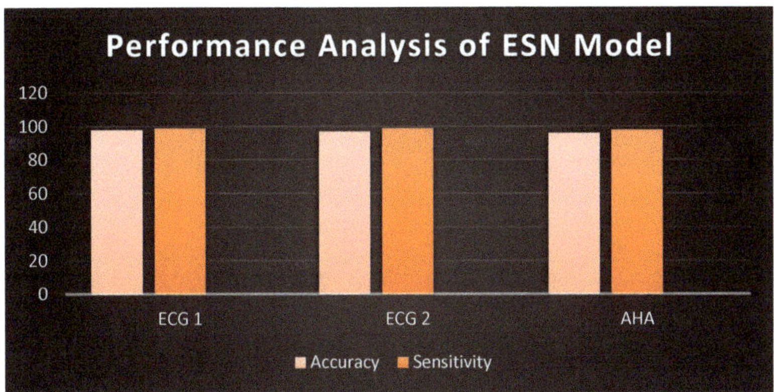

Figure 11. Graphical representation of the performance Analysis of the ESN model.

It is noticed that the complexity of RC is higher in random connections, whereas the cycle-based ESN model has low computational cost and it is suitable for implementation in real time scenarios. In the learning phase, this study used Ridge regression to optimize the desired output weights and avoid overfitting, whereas in the testing phase, it only involves sparse connectivity which helps to reduce the number of heavy multiplications and additions. In this study, all experimental work was conducted using MATLAB (2018a) on a desktop computer running with 128 GB RAM and 16 cores. Once all scaling parameters were adjusted in the model, the system only required eight minutes in training with a 30 min duration ECG dataset. However, the testing runtime was observed that it takes 1.5 s for feature extraction or selection from ECG beats and the model takes 5 s to classify the heartbeat either as (N + SVB) or (A + VB). The total approximation time is estimated as 2.0 s and total time complexity is observed $O(F(n)^2 E)$ where E denotes the training examples and F(n) is the number of features.

In contrast, this study examined the limitation of our proposed ESN in implementation and the finding is that the classifier needs to wait until the reference beat is computed. Once the template beat is computed, the system is ready to implement the real-time beats for classification. The system is able to compute the new template beat when more than 10 beats were classified properly and the model starts retraining by using the new template beat and it only consumes 2.0 s for the training model.

3.3. Comparison with State-of-the-art Methods

In the literature [12,16,54,55], the author noticed that many studies used the MIT-BIH arrhythmia dataset for the heartbeat classification, whereas this study used different datasets. Therefore, it is obvious that this study cannot present a fair comparison between the current study and state-of-the-art methods. To support a fair comparison analysis, this study used the MIT-BIH arrhythmia dataset for heartbeat classification using our proposed non-random ESN model. This method is focused on a single lead and ventricular heartbeat classification. Therefore, this technique considered only two classes: (N + SVB) or (A + VB). For MLII and V1, both leads were taken individually to classify the ventricular heartbeats from the ECG signals. This methodology outperforms the rest of the other studies that were discussed in Table 4. Moreover, some studies in the literature only focused on one lead classification and the computational cost of the proposed algorithm was not suitable for real-time application [9,56]. However, this study used both leads for classification and the technique computational cost and performance of our technique is noteworthy for real-time implementation. Thus, the author compared this study with those state-of-the-art methods that used both leads for learning the algorithm and the results were extracted from their confusion matrix which is only related to ventricular heartbeats.

Table 4. Comparison analysis of the proposed work with existing studies.

Existing Studies	Leads	+P (%)	Se (%)
[57]	MLII (modified lead II) and V1	92.8	85.5
[58]	MLII and V1	87.0	89.0
[17]	MLII and V1	98.0	91.8
Proposed Work	MLII	98.98	98.98
	V1	98.96	97.95

4. Conclusions

The proposed algorithm is especially designed for implementation in medical wearable wireless gadgets as it is fast with less power consumption. The study's proposed methodology used single ECG lead signals to classify the ventricular heartbeats. Hence, the performance of the proposed classifier is compared to other existing methodologies. The system provides a significant contribution in the field of MIoT (Medical Internet of Things) and also provides the ability to train the new dataset for the enhancement of system performance. The quality of the system is capable enough to be implemented in wearable wireless devices or MIoT gadgets. In future work, there will be the need to modify this algorithm for other annotated heartbeat classification.

Author Contributions: All authors contributed equally to this work.

Funding: This work was supported by the University of Malaya Research BKP Special Grant BKS006-2018.

Acknowledgments: The authors would like to thank anonymous reviewers for their valuable comments and support.

References

1. Writing Group Members; Thom, T.; Haase, N.; Rosamond, W.; Howard, V.J.; Rumsfeld, J.; Manolio, T.; Zheng, Z.J.; Flegal, K.; O'Donnell, C.; et al. Heart disease and stroke statistics—2006 update: A report from the American Heart Association Statistics Committee and Stroke Statistics Subcommittee. *Circulation* **2006**, *113*, e85–e151. [PubMed]
2. Stevens, G.; Mascarenhas, M.; Mathers, C. WHO brochure». *Bull. World Health Organ.* **2009**, *87*, 646. [CrossRef] [PubMed]
3. Li, H.; Ge, J. Cardiovascular diseases in China: Current status and future perspectives. *IJC Heart Vasc.* **2015**, *6*, 25–31. [CrossRef] [PubMed]
4. Carnevale, L.; Celesti, A.; Fazio, M.; Bramanti, P.; Villari, M. Heart disorder detection with menard algorithm on apache spark. In *European Conference on Service-Oriented and Cloud Computing*; Springer: Berlin/Heidelberg, Germany, 2017; pp. 229–237.
5. Melillo, P.; Castaldo, R.; Sannino, G.; Orrico, A.; de Pietro, G.; Pecchia, L. Wearable technology and ECG processing for fall risk assessment, prevention and detection. In Proceedings of the 2015 37th Annual International Conference of the IEEE, Engineering in Medicine and Biology Society (EMBC), Milan, Italy, 25–29 August 2015; pp. 7740–7743.
6. Sannino, G.; de Pietro, G. An evolved ehealth monitoring system for a nuclear medicine department. In Proceedings of the Developments in E-Systems Engineering (DeSE), Dubai, United Arab Emirates, 6–8 December 2011; pp. 3–6.
7. Abbasi, Q.H.; Rehman, M.U.; Qaraqe, K.; Alomainy, A. Advances in body-centric wireless communication: Applications and state-of-the-art. *Institution of Engineering and Technology* **2016**.
8. De Chazal, P.; O'Dwyer, M.; Reilly, R.B.R. Automatic classification of heartbeats using ECG morphology and heartbeat interval features. *IEEE Trans. Biomed. Eng.* **2004**, *51*, 1196–1206. [CrossRef] [PubMed]
9. Tsipouras, M.G.; Fotiadis, D.I. Automatic arrhythmia detection based on time and time–frequency analysis of heart rate variability. *Comput. Methods Progr. Biomed.* **2004**, *74*, 95–108. [CrossRef]

10. Alonso-Atienza, F.; Morgado, E.; Fernandez-Martinez, L.; Garcia-Alberola, A.; Rojo-Alvarez, J.L. Detection of life-threatening arrhythmias using feature selection and support vector machines. *IEEE Trans. Biomed. Eng.* **2014**, *61*, 832–840. [CrossRef]

11. Zhang, Z.; Dong, J.; Luo, X.; Choi, K.S.; Wu, X. Heartbeat classification using disease-specific feature selection. *Comput. Biol. Med.* **2014**, *46*, 79–89. [CrossRef]

12. Huang, H.; Liu, J.; Zhu, Q.; Wang, R.; Hu, G. A new hierarchical method for inter-patient heartbeat classification using random projections and RR intervals. *Biomed. Eng. Online* **2014**, *13*, 90. [CrossRef]

13. Zidelmal, Z.; Amirou, A.; Ould-Abdeslam, D.; Merckle, J. ECG beat classification using a cost sensitive classifier. *Comput. Methods Progr. Biomed.* **2013**, *111*, 570–577. [CrossRef]

14. Garcia, G.; Moreira, G.; Menotti, D.; Luz, E.J.S.R. Inter-Patient ECG Heartbeat Classification with Temporal VCG Optimized by PSO. *Sci. Rep.* **2017**, *7*, 10543. [CrossRef] [PubMed]

15. Qurraie, S.S.; Afkhami, R.G.J.B.E.L. ECG arrhythmia classification using time frequency distribution techniques. *International journal of Intelligent Engineering and Systems* **2017**, *7*, 325–332.

16. Ye, C.; Kumar, B.V.; Coimbra, M.T. Heartbeat classification using morphological and dynamic features of ECG signals. *IEEE Trans. Biomed. Eng.* **2012**, *59*, 2930–2941. [PubMed]

17. Ye, C.; Kumar, B.V.; Coimbra, M.T. An Automatic Subject-Adaptable Heartbeat Classifier Based on Multiview Learning. *IEEE J. Biomed. Health Inform.* **2016**, *20*, 1485–1492. [CrossRef] [PubMed]

18. Mar, T.; Zaunseder, S.; Martínez, J.P.; Llamedo, M.; Poll, R. Optimization of ECG classification by means of feature selection. *IEEE Trans. Biomed. Eng.* **2011**, *58*, 2168–2177. [CrossRef] [PubMed]

19. Elhaj, F.A.; Salim, N.; Harris, A.R.; Swee, T.T.; Ahmed, T. Arrhythmia recognition and classification using combined linear and nonlinear features of ECG signals. *Comput. Methods Progr. Biomed.* **2016**, *127*, 52–63. [CrossRef] [PubMed]

20. Raj, S.; Ray, K.C.; Shankar, O. Cardiac arrhythmia beat classification using DOST and PSO tuned SVM. *Comput. Methods Progr. Biomed.* **2016**, *136*, 163–177. [CrossRef]

21. Castillo, O.; Melin, P.; Ramírez, E.; Soria, J. Hybrid intelligent system for cardiac arrhythmia classification with Fuzzy K-Nearest Neighbors and neural networks combined with a fuzzy system. *Expert Syst. Appl.* **2012**, *39*, 2947–2955. [CrossRef]

22. Saini, I.; Singh, D.; Khosla, A. QRS detection using K-Nearest Neighbor algorithm (KNN) and evaluation on standard ECG databases. *J. Adv. Res.* **2013**, *4*, 331–344. [CrossRef]

23. Dokur, Z.; Ölmez, T. ECG beat classification by a novel hybrid neural network. *Comput. Methods Progr. Biomed.* **2001**, *66*, 167–181. [CrossRef]

24. Martis, R.J.; Acharya, U.R.; Min, L.C. ECG beat classification using PCA, LDA, ICA and discrete wavelet transform. *Biomed. Signal Process. Control* **2013**, *8*, 437–448. [CrossRef]

25. Inan, O.T.; Giovangrandi, L.; Kovacs, G.T. Robust neural-network-based classification of premature ventricular contractions using wavelet transform and timing interval features. *IEEE Trans. Biomed. Eng.* **2006**, *53*, 2507–2515. [CrossRef] [PubMed]

26. Javadi, M.; Ebrahimpour, R.; Sajedin, A.; Faridi, S.; Zakernejad, S. Improving ECG classification accuracy using an ensemble of neural network modules. *PLoS ONE* **2011**, *6*, e24386. [CrossRef] [PubMed]

27. Afkhami, R.G.; Azarnia, G.; Tinati, M.A. Cardiac arrhythmia classification using statistical and mixture modeling features of ECG signals. *Pattern Recognit. Lett.* **2016**, *70*, 45–51. [CrossRef]

28. Wu, Z.; Ding, X.; Zhang, G. A novel method for classification of ECG arrhythmias using deep belief networks. *J. Comput. Intell. Appl.* **2016**, *15*, 1650021. [CrossRef]

29. Acharya, U.R.; Oh, S.L.; Hagiwara, Y.; Tan, J.H.; Adam, M.; Gertych, A.; San Tan, R. A deep convolutional neural network model to classify heartbeats. *Comput. Biol. Med.* **2017**, *89*, 389–396. [CrossRef] [PubMed]

30. Kiranyaz, S.; Ince, T.; Gabbouj, M. Real-time patient-specific ECG classification by 1-D convolutional neural networks. *IEEE Trans. Biomed. Eng.* **2016**, *63*, 664–675. [CrossRef]

31. Krasteva, V.; Jekova, I. QRS template matching for recognition of ventricular ectopic beats. *Ann. Biomed. Eng.* **2007**, *35*, 2065–2076. [CrossRef]

32. Jaeger, H.J.B. *The "Echo State" Approach to Analysing and Training Recurrent Neural Networks-with An Erratum Note*; German National Research Center for Information Technology GMD Technical Report; German National Research Center: Bonn, Germany, 2001; Volume 148, p. 13.

33. Lukoševičius, M.; Jaeger, H. Reservoir computing approaches to recurrent neural network training. *Comput. Sci. Rev.* **2009**, *3*, 127–149. [CrossRef]

34. Greenwald, S.D.; Patil, R.S.; Mark, R.G. Improved detection and classification of arrhythmias in noise-corrupted electrocardiograms using contextual information. In Proceedings of the Proceedings Computers in Cardiology, Chicago, IL, USA, 23–26 September 1990; pp. 461–464.

35. Goldberger, A.L.; Amaral, L.A.; Glass, L.; Hausdorff, J.M.; Ivanov, P.C.; Mark, R.G.; Mietus, J.E.; Moody, G.B.; Peng, C.-K.; Stanley, H.E.J.C. PhysioBank, PhysioToolkit, and PhysioNet: Components of a new research resource for complex physiologic signals. *Circulation* **2000**, *101*, e215–e220. [CrossRef]

36. ANSI/AAMI EC57:1998. *Testing and Reporting Performance Results of Cardiac Rhythm and ST Segment Measurement Algorithms*; AAMI: Brisbane, Australia, 1998.

37. Friesen, G.M.; Jannett, T.C.; Jadallah, M.A.; Yates, S.L.; Quint, S.R.; Nagle, H.T. A comparison of the noise sensitivity of nine QRS detection algorithms. *IEEE Trans. Biomed. Eng.* **1990**, *37*, 85–98. [CrossRef] [PubMed]

38. Chandrakar, B.; Yadav, O.P.; Chandra, V.K. A survey of noise removal techniques for ECG signals. *Int. J. Adv. Res. Comput. Commun. Eng.* **2013**, *2*, 1354–1357.

39. Cuomo, S.; De Pietro, G.; Farina, R.; Galletti, A.; Sannino, G. A revised scheme for real time ecg signal denoising based on recursive filtering. *Biomed. Signal Process. Control* **2016**, *27*, 134–144. [CrossRef]

40. Farina, R.; Dobricic, S.; Storto, A.; Masina, S.; Cuomo, S. A revised scheme to compute horizontal covariances in an oceanographic 3D-VAR assimilation system. *J. Comput. Phys.* **2015**, *284*, 631–647. [CrossRef]

41. De Chazal, P.; Reilly, R.B. A patient-adapting heartbeat classifier using ECG morphology and heartbeat interval features. *IEEE Trans. Biomed. Eng.* **2006**, *53*, 2535–2543. [CrossRef] [PubMed]

42. Pratt, W.K. *Digital Image Processing: PIKS Scientific Inside*; Wiley-Interscience: Hoboken, NJ, USA, 2007.

43. IEEE Acoustics, Speech, and Signal Processing Society. Digital Signal Processing Committee. In *Programs for digital signal processing*; IEEE: Piscataway, NJ, USA, 1979.

44. Pan, J.; Tompkins, W.J. A real-time QRS detection algorithm. *IEEE Trans. Biomed. Eng.* **1985**, *32*, 230–236. [CrossRef] [PubMed]

45. Acharya, U.R.; Fujita, H.; Oh, S.L.; Raghavendra, U.; Tan, J.H.; Adam, M.; Gertych, A.; Hagiwara, Y. Automated identification of shockable and non-shockable life-threatening ventricular arrhythmias using convolutional neural network. *Future Gen. Comput. Syst.* **2018**, *79*, 952–959. [CrossRef]

46. Nong, W. A novel algorithm for ventricular arrhythmia classification using a fuzzy logic approach. *Aust. Phys. Eng. Sci. Med.* **2016**, *39*, 903–912.

47. Jain, A.K.; Duin, R.P.; Mao, J. Statistical pattern recognition: A review. *IEEE Trans. Pattern Anal. Mach. Intell.* **2000**, *22*, 4–37. [CrossRef]

48. Kim, Y.J.; Heo, J.; Park, K.S.; Kim, S. Proposition of novel classification approach and features for improved real-time arrhythmia monitoring. *Comput. Biol. Med.* **2016**, *75*, 190–202. [CrossRef]

49. Maass, W.; Natschläger, T.; Markram, H. Real-time computing without stable states: A new framework for neural computation based on perturbations. *Neural Comput.* **2002**, *14*, 2531–2560. [CrossRef] [PubMed]

50. Yamazaki, T.; Tanaka, S. The cerebellum as a liquid state machine. *Neural Netw.* **2007**, *20*, 290–297. [CrossRef] [PubMed]

51. Brunner, D.; Soriano, M.C.; Mirasso, C.R.; Fischer, I. Parallel photonic information processing at gigabyte per second data rates using transient states. *Nat. Commun.* **2013**, *4*, 1364. [CrossRef] [PubMed]

52. Albert, A. *Regression and the Moore-Penrose Pseudoinverse*; Elsevier: Amsterdam, The Netherlands, 1972.

53. Tikhonov, A.; Goncharsky, A.; Stepanov, V.; Yagola, A.G. Numerical methods for the solution of ill-posed problems (Mathematics and its Applications). 1995.

54. Kutlu, Y.; Kuntalp, D. Feature extraction for ECG heartbeats using higher order statistics of WPD coefficients. *Comput. Methods Progr. Biomed.* **2012**, *105*, 257–267. [CrossRef] [PubMed]

55. Zhang, H. System for Cardiac Arrhythmia Detection and Characterization. U.S. Google Patent 8,233,972, 31 July 2012.

56. Teijeiro, T.; Félix, P.; Presedo, J.; Castro, D. Heartbeat classification using abstract features from the abductive interpretation of the ECG. *IEEE J. Biomed. Health Inform.* **2018**, *22*, 409–420. [CrossRef]

57. Llamedo, M.; Martínez, J.P. Heartbeat classification using feature selection driven by database generalization criteria. *IEEE Trans. Biomed. Eng.* **2011**, *58*, 616–625. [CrossRef]
58. Llamedo, M.; Martínez, J.P. An automatic patient-adapted ECG heartbeat classifier allowing expert assistance. *IEEE Trans. Biomed. Eng.* **2012**, *59*, 2312–2320. [CrossRef]

Article

A Real Time and Lossless Encoding Scheme for Patch Electrocardiogram Monitors

Hong-Wen Fang [1] and Chih-Cheng Lu [1,2,3,*]

[1] Graduate Institute of Mechanical and Electrical Engineering, National Taipei University of Technology, Taipei 106, Taiwan; alan.fhw@gmail.com

[2] Graduate Institute of Mechatronic Engineering, National Taipei University of Technology, Taipei 106, Taiwan

[3] Department of Mechanical Engineering, National Taipei University of Technology, Taipei 106, Taiwan

* Correspondence: cclu23@ntut.edu.tw

Received: 30 October 2018; Accepted: 20 November 2018; Published: 24 November 2018

Abstract: Cardiovascular diseases are the leading cause of death worldwide. Due to advancements facilitating the integration of electric and adhesive technologies, long-term patch electrocardiogram (ECG) monitors (PEMs) are currently used to conduct daily continuous cardiac function assessments. This paper presents an ECG encoding scheme for joint lossless data compression and heartbeat detection to minimize the circuit footprint size and power consumption of a PEM. The proposed encoding scheme supports two operation modes: fixed-block mode and dynamic-block mode. Both modes compress ECG data losslessly, but only dynamic-block mode supports the heartbeat detection feature. The whole encoding scheme was implemented on a C-platform and tested with ECG data from MIT/BIH arrhythmia databases. A compression ratio of 2.1 could be achieved with a normal heartbeat. Dynamic-block mode provides heartbeat detection accuracy at a rate higher than 98%. Fixed-block mode was also implemented on the field-programmable gate array, and could be used as a chip for using analog-to-digital convertor-ready signals as an operation clock.

Keywords: electrocardiography (ECG); patch ECG monitor

---

1. Introduction

According to the World Health Organization, cardiovascular diseases (CVD) are the leading cause of death worldwide [1]. An estimated 17.7 million people died from CVD in 2015, representing 31% of all global deaths. Of these, an estimated 7.4 million and 6.7 million deaths were caused by coronary heart disease and stroke, respectively.

Of the 16 million people aged under 70 years who died from non-communicable diseases in 2011, 37% of these deaths were caused by CVD, and 82% of them lived in low and middle-income countries, where people often do not have access to integrated primary health care programs for the early detection and treatment of such illnesses. The cost of CVD treatment escalates quickly with the rapidly increasing requirements for supervision and medical management, and traditional health care infrastructures are easily overwhelmed by the demands of effective late-stage treatment. However, such costs can potentially be reduced by using systems that monitor individuals in the course of their daily activities, which would thus reduce the need for inpatient care or visits to the primary physician.

Early detection and prevention are critical for reducing both the direct and indirect costs of CVD, and researchers have responded to this need by focusing on improving techniques that assess cardiac function, including electrocardiogram (ECG) analysis, which helps screen for various cardiac abnormalities. Current ECG monitoring systems provide continuous, simple, risk-free, and inexpensive recordings of the heart's electrical and muscular functions [2], and advances in technology have changed the way that ECG signals are collected, stored, and analyzed to realize "proactive health care". This has enabled continuous health care monitoring through the use of smartphones [3–7], because

ECG signals can be easily transmitted through common communication channels (such as phone lines or wireless channels). In addition, wireless body sensor networks (WBSNs) [8,9] promise large-scale solutions and can be used in place of mobile phones. These solutions automatically report cardiac signals to health care providers, which make them extremely useful in ambulatory settings.

To minimize power consumption and enable long-term ECG data logging, a Holter monitor is used to record patients' ECG data offline continuously for 24–72 h. However, the Holter device does not analyze data or detect a disease, and it does not provide calculated diagnosis information to the hospital, doctor, or patient in response to a critical heart condition. Nevertheless, it has emerged as the most common ECG battery-operated platform that does not have cellular or WBSN connections [10]. Advances in semiconductor technology have enabled additional ECG signal storage with even lower power consumption; these ECG signals provide information that is subsequently used by clinicians to determine the mechanisms underlying CVD, and this development is expected to ultimately lead to effective treatment.

The emerging trend is to use wearable devices to record patients' ECG data [11]. However, the Holter device typically has six to 10 sensors that must be arrayed on the body, which makes it inconvenient for everyday use. Recent technological advances have thus produced single-use, wearable, monitoring devices that are capable of continuously recording ECG signals for a period of three days or more. A wearable patch ECG monitor (PEM) records ECG signals using an embedded electronic circuit that is attached to a patient's chest with invisible electrodes and lead wires. The device can be considered a wearable health care system that is based on the use of knitted integrated sensors connected to electrodes using a conductive and piezoresistive yarn [12]. PEMs are typically highly energy-efficient and provide an analog readout application-specific integrated circuit (ASIC) for signal acquisition, amplification, and analog-to-digital conversion. They also have the advantage of using wireless network interfaces instead of power-hungry wireless links while simultaneously maintaining an extremely small footprint.

ECG monitoring systems produce large volumes of data that must be compressed for efficient processing, storage, and transmission. Many wavelet transform–based methods have been proposed for use with the lossy data compression technique [13–23], and these have fine visual qualities that minimize reconstruction errors. Such methods achieve high compression ratios (CRs) and have no diagnostic-affecting features relating to reconstructed signals. Lossless compression schemes [24,25] employ original signals that are not reconstructed, and because no signal distortion can occur with their use, they are preferable for diagnostic applications. In this respect, Koski [26] used LZ77, complex extraction, and Huffman coding to achieve lossless ECG compression; Giurcâneanu et al. [27] proposed a lossless compression scheme using contexts and R–R interval estimation; and Miaou and Chao [28] proposed a lossless compression method using vector quantization and wavelet transforms.

The encoding scheme proposed in this study addresses real-time application under resource-limited conditions for field-programmable gate array (FPGA), ASIC, or chip implementation, and the primary objective of this design was to achieve easy circuit integration (Figure 1). The ECG signal-acquiring module uses an Analogy-to-Digital converter (ADC) chip to convert an analogy signal to digital raw data for processing. This implementation requires less modification when inserting the proposed chip between the ADC chip and the processor, because it uses the existing sample clock from the ADC data-ready signal as the only operation clock. This chip encodes raw data from the ADC output, and sends it for processing using the same connection interface. The encoding scheme comprises an arithmetic logic unit, such as subtraction or rotation, and a counter unit for minimal power consumption. A PEM uses this encoding scheme by employing the existing ASIC at no additional cost. This approach offers two operation modes: fixed-block mode and dynamic-block mode. Fixed-block mode has one prefix parameter, a "split resolution", which affects the CR and was experimentally applied to the MIT/BIH Arrhythmia Database [29] to identify parameters for optimal compression. Dynamic-block mode uses the selected "split resolution" from the fixed-block mode to compress ECG data while simultaneously producing a quantity of continue flat data, which could be as heartbeat detection. The final CR exceeds 2.1; however, and the heartbeat detection accuracy reaches 98% with a normal beat.

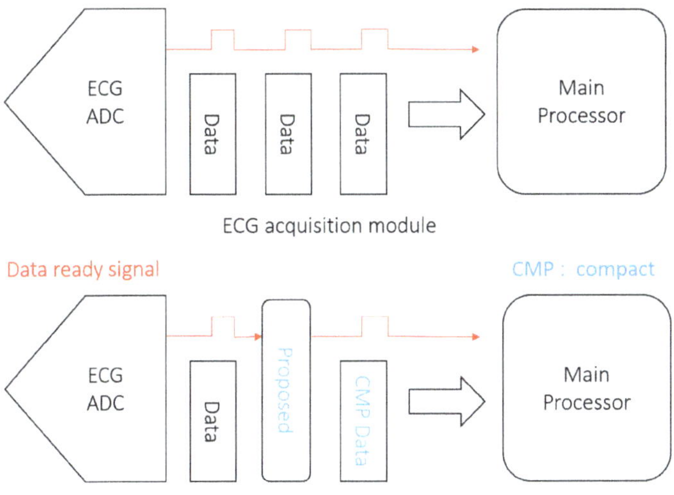

Figure 1. Easy chip integration.

2. Design, Materials, and Methods

ECG signals comprise low-frequency (P and T waves and ST segments) and medium-frequency (QRS complexes) components [30]. This study aimed to reduce the storage space required for low-frequency segmentation using the fixed and dynamic-block modes. In this study, "block" refers to a quantity of ECG records, and it is the basic unit of data to be compressed. Both modes are used for real-time and lossless ECG data compression for wearable devices, particularly for PEMs. Herein, fixed-block mode is discussed in three main sections: Section 2.1 discusses delta coding [31], Section 2.2 discusses zero separation, and Section 2.3 discusses zero run length encoding (ZRLE). In Section 2.4, the format output provides a detailed description of the compressed result. Since the dynamic-block mode is similar to fixed-block mode, Section 2.5 explains only the major differences between the two modes.

Figure 2 shows a block diagram of the process used in the proposed encoding scheme. This encoding scheme processes only the voltage values. First, delta coding is used to generate delta values by reducing the dynamic range of the ECG volume. Next, the delta value is rotated one bit to the left in zero separation. This rotation can displace the sign flag of the highest bit, which produces the lowest bit so that the remaining bits shift to the left. The rotated delta value is then split into low bits and high bits with the prefix parameter "split resolution". Finally, one of the listed coding methods, ZRLE, which is based on run length encoding (RLE), is used to encode the zero value of high-bit data only. This scheme bypasses low-bit data and can be subsequently enhanced.

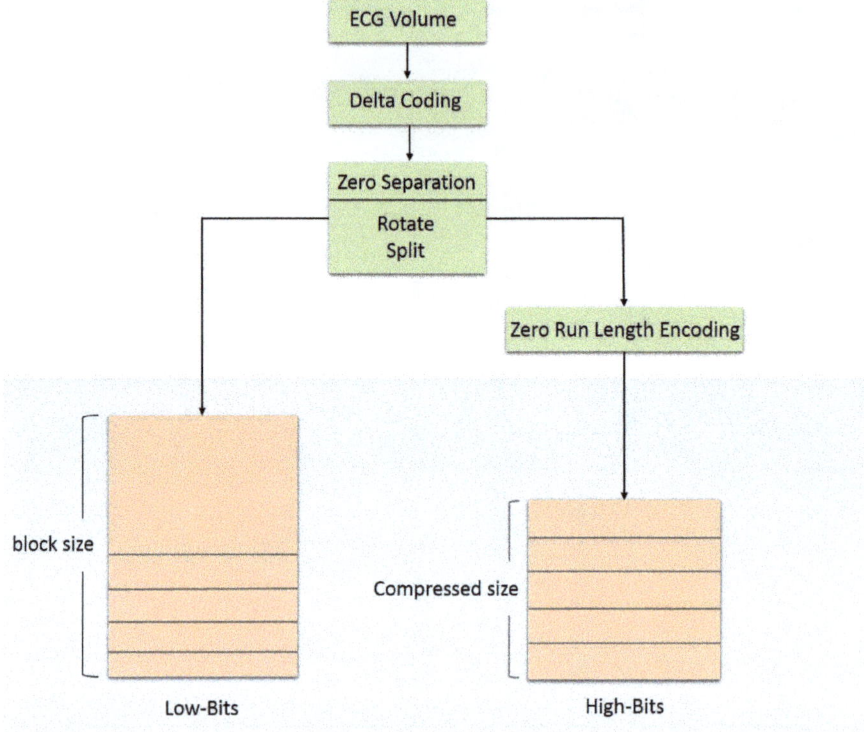

Figure 2. Flow chart.

2.1. Delta Coding

Delta coding is applied to reduce the dynamic range of original ECG signals. Subsequent encoded samples are generated from the difference between the current sample and the previous sample of the original ECG signal. An ECG signal contains the shape and size of the P–QRS–T and U waves, and time intervals between various peaks. Each interval contains different frequencies and variations, and the dynamic range of the delta coding volume therefore depends on the input waves. The minimal dynamic range is the P–Q or T–U wave, and the maximal dynamic range is the QRS complex.

A 30-minute ECG recording of MIT/BIH 100 was taken from the MIT/BIH database, which uses an 11-bit resolution over a 10-mv range, and has a dynamic signal range from 485 (2.3 mv) to 1308 (6.3 mv). In addition to the first recorded value of 995 that has a previous sample of 0, delta coding generates a new dynamic range with a delta volume from −217 (−1.05 mv) to 141 (0.688 mv). Since MIT/BIH has a resolution of 11 bits, the first delta value of the first sampled ECG volume can be calculated using the initial previous value of 1024 (middle value) to reduce the dynamic range of the delta volume. Figure 3 compares the MIT/BIH 100 probability of occurrence of each distinct symbol between the ECG (original) volume and delta volume variance; the distinct delta volume variance is more centralized than the original volume. The dynamic range of the delta volume is limited to between −217 and 141, which can be represented by nine bits (in addition to the first record of 995). Figure 3 shows the MIT/BIH 100 ECG volume and delta volume.

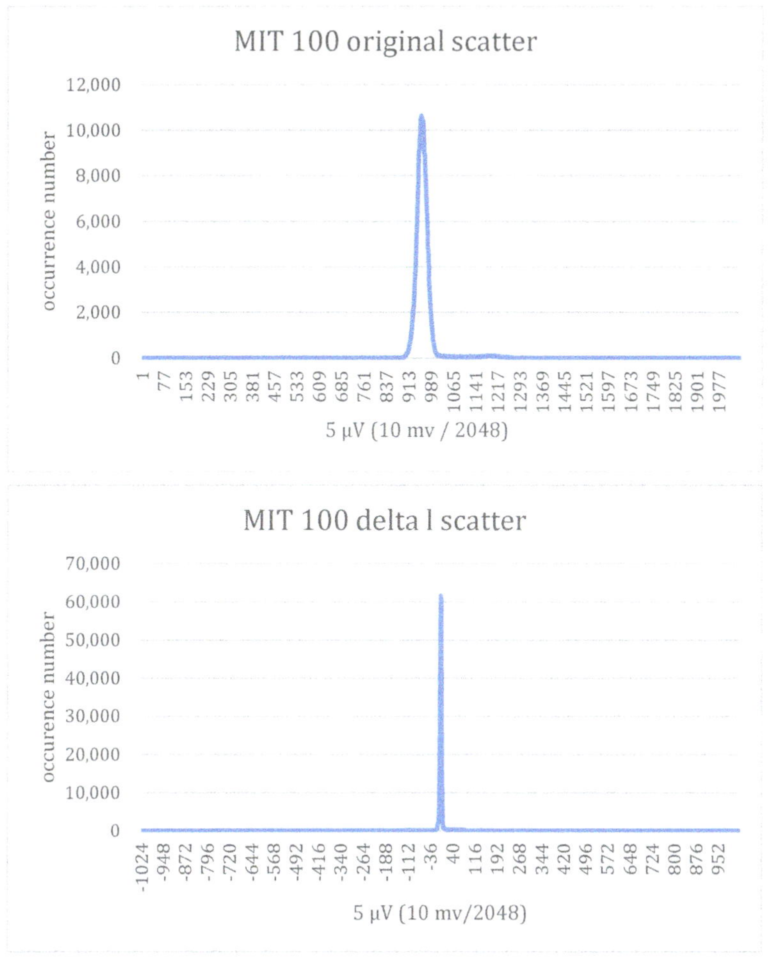

Figure 3. Comparison of probability of occurrence of each distinct symbol for an electrocardiogram (ECG) (original) volume and delta volume variance with MIT/BIH Arrhythmia Database 100.

2.2. Zero Separation

Since an ECG is composed of low-frequency (P and T waves and ST segments) and medium-frequency (QRS complexes) components, each ECG interval can be stored in different resolutions. Figure 4 shows that a nine-bit resolution can be used to present delta sequence data in MIT/BIH case 100. This step splits the delta value into two parts: high bits and low bits. Ideally, high bits contain only zero values with specific bit quantities (resolutions) in each interval. With a low-frequency component, a flat signal means a smaller dynamic range of delta volume and produces a zero value with a higher resolution in high bits, and a sign flag problem arises in this step when the delta value is negative. One additional bit, the Most Significant Bit (MBS), increases the data size to indicate either a positive or a negative delta value, and results in many non-zero values in high bits. Since neither maximal (2048) nor minimal (0) ECG signal values appear nearby, the additional bit can be ignored. The delta value must be rotated one bit to the left to encode the bit with the most negative value as zero. Figure 5 shows the MIT/BIH recording of 100 original volumes and rotated delta volumes.

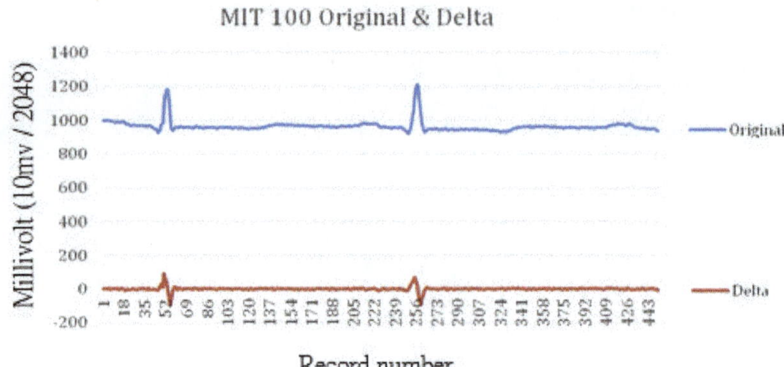

Figure 4. MIT/BIH Arrhythmia Database recording 100 original ECG volumes and delta volumes.

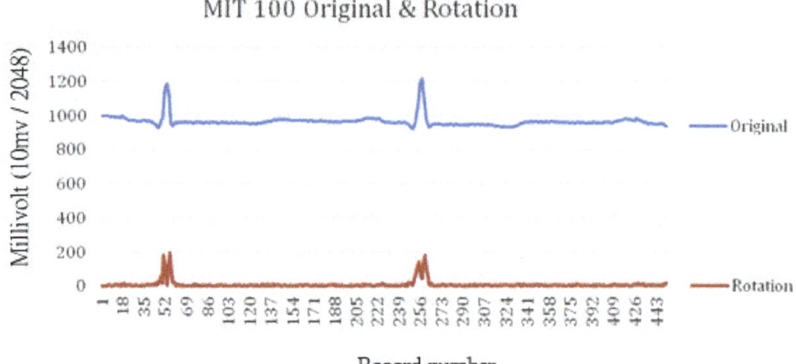

Figure 5. MIT/BIH Arrhythmia Database recording 100 ECG volumes and rotated delta volumes.

2.3. ZRLE

Preprocessing with delta coding for optimal selective Huffman coding is used to achieve lossless ECG compression [32]. Huffman coding creates variable-length codes with shorter code words for higher probabilities, and each is represented by an integer bit number. Huffman coding is relatively simple, and it is the best coding scheme possible when coded words are restricted to integer lengths [32,33]. Section 2.2 describes how zero separation increases the zero-value probability of high bits, and this can then be compressed using Huffman. For real-time encoding and minimal implementation complexity, a new method based on RLE [33] was designed with a high probability of zero bases. RLE is a simple technique that is used for digital data compression and represents successive runs of the same value in the data when the value is followed by a counter, rather than the original run of values. ZRLE is modified from RLE to encode zero values only. This method replaces continual zero values as one zero and a number of consecutive zeros, and it requires a quantity of consecutive zeros greater than two. In addition, it fails to compress a single zero with a pair of 0 and 1. The resolution of high bits defines the maximum number of consecutive zeros. For example, four-bit high-bit resolution presents a maximum number of consecutive zeros of 16, or two to the fourth power. However, a lower resolution corresponds with more consecutive zeros in high bits, but an increasing low-bits resolution reduces the total CR. Although a QRS complex is of medium-frequency and has a higher dynamic data range, it provides fewer periods in a single heartbeat. Since it is not possible to know the encoded size of high bits, two consecutive zeros represent the end tag that identifies the end of one block. When a block of data is encoded, the end tag is appended to the last record (that is not a pair of zeros) and the counter. Thus, the decoding process restores the high-bit volume with the same predefined block size, or the zero value is used to make up for the shortage of records in a high-bit volume. The end tag affects the CR terminating the data string with 0, 0, and a suitable segmentation has an end tag to eliminate longer continual zeros in the tail. This experiment aimed to provide a high-bit resolution encoding scheme to achieve the optimal CR for use in dynamic-block mode.

2.4. Format Output

ZRLE compresses high-bit data produced by delta coding and zero separation. Zero separation splits delta data with a selected resolution that must be experimentally identified. Figure 2 shows a flow chart where the low-bit and high-bit streams are the compressed results of continuous, real-time ECG data. Zero separation directly transfers the low-bit portion of the rotated delta data into the low-bit stream without compression. The maximal CR is then determined by the high-bit resolution and limited (low-bit resolution + high-bit resolution)/(low-bit resolution). Figure 6 shows the encoded data format of a fixed-block size, where the M value is determined by ZRLE, and should be smaller than N, the block size.

Figure 6. Fixed-block mode output format.

2.5. Dynamic-Block Mode and Heartbeat

The heartbeat ratio and sampling rate influence the quantity of samples for each interval, such as P, PR segments, QRS complexes, ST segments, T, U, and V in fixed-block mode, and an increment counter counts the quantity of the encoded data and appends an end tag to the high-bit output stream. The optimal CR has a valid end tag when using ECG data block segmentation with QRS complexes, and the sequence value following the QRS complexes in one heartbeat can be replaced by an end tag. Although many real-time QRS detection algorithms have been proposed, dynamic-block mode can simply modify the counter for QRS complex detection. In addition, counting continual zeros in high-bits can be used roughly to segment the ECG stream by QRS complexes. When the quantity of consecutive zeros exceeds the threshold, one-third of the sampling rate is calculated in the normal heartbeat range from 0.8–3 HZ [34]. The next non-zero has a high probability of being a QRS complex, which will thus determine the block size of the dynamic-block mode. The compression procedure is identical to that of the fixed-block mode; the only major difference is the block size K, which is generated through non-QRS-complex segmentation, and is inserted in the initial output (Figure 4). Figure 7 shows the output format for dynamic-block mode.

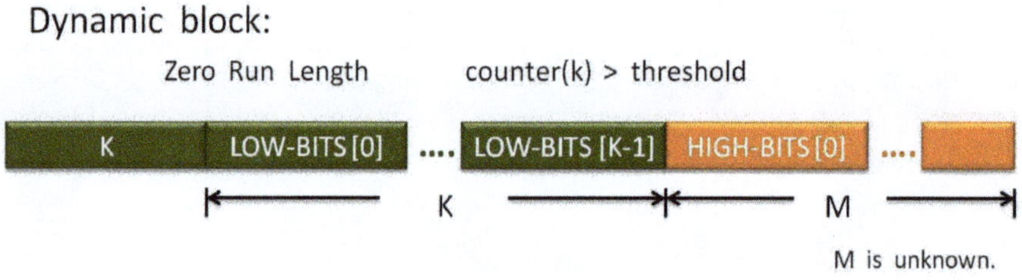

Figure 7. Dynamic-block mode output format.

3. Results

This encoding scheme was implemented in C using GCC 4.01 on FreeBSD 10.1 and tested using the MIT/BIH Arrhythmia Database records 100 to 109. Each record had a sampling rate of 360 Hz and 11-bit resolution. The CR calculation is defined as follows, and the results are provided in the following sections.

$$CR = \text{Number of bits in}/\text{Number of bits out}$$

3.1. High-Bit Resolution

With respect to the fixed-block mode, this research reserved 512 records as the block size and 512 units of output buffer space to identify three to eight suitable high-bit resolutions. The CRs are presented in Table 1 and Figure 8. The preferred CR was found at a resolution of seven bits on recording 100. Figure 9 shows the best case for high-bit and low-bit results for recording 100. Recording 107 provided the lowest performance of all 10 records because it contained an ECG signal with an abnormal "paused" heartbeat. The best CR of recording 107 was at a resolution of five bits for high bits, and it was the only recording that provided a high performance at five bits. Figure 10 presents the original ECG signal, which has a longer QRS complex than a standard ECG signal does. Since the QRS complex has a high-variation signal, high bits are used to represent the value and have short continual zeros. In contrast, zero separation exhibited a less favorable performance.

Table 1. Compression rate for high-bit resolution.

	3 bits	4 bits	5 bits	6 bits	7 bits	8 bits
100	1.342	1.453	1.679	1.981	2.220	1.637
101	1.351	1.457	1.693	1.973	2.172	1.455
102	1.364	1.464	1.713	1.971	2.177	1.509
103	1.263	1.445	1.679	1.936	2.119	1.412
104	1.350	1.459	1.680	1.889	1.936	1.417
105	1.370	1.501	1.644	1.814	1.798	1.234
106	1.307	1.468	1.686	1.922	1.742	1.100
107	1.263	1.458	1.586	1.576	1.341	1.090
108	1.371	1.552	1.733	1.923	1.749	1.215
109	1.366	1.494	1.609	1.822	1.824	1.276

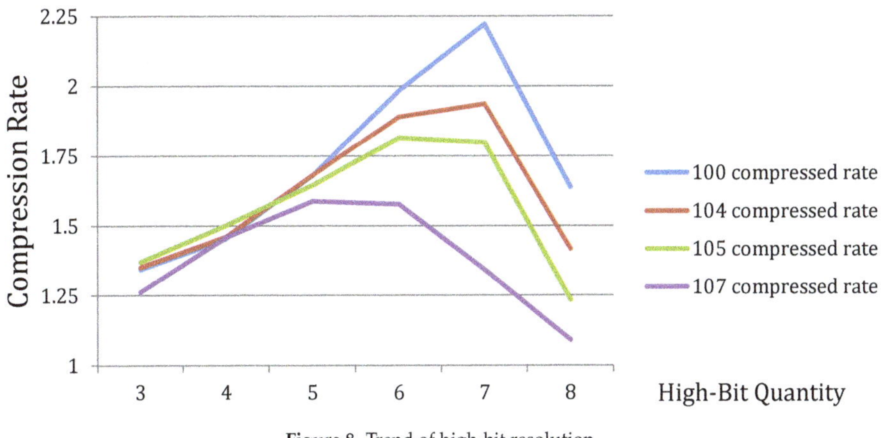

Figure 8. Trend of high-bit resolution.

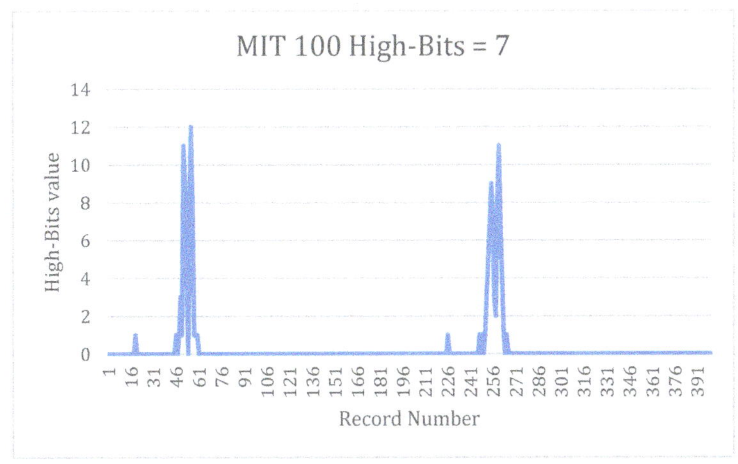

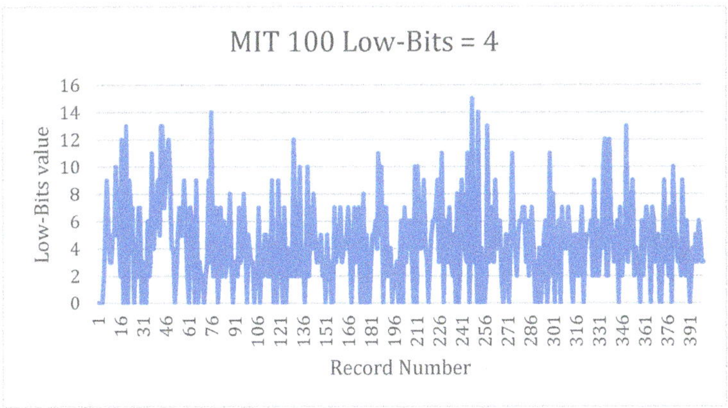

Figure 9. Best case: MIT/BIH Arrhythmia Database recording 100 high bits = 7.

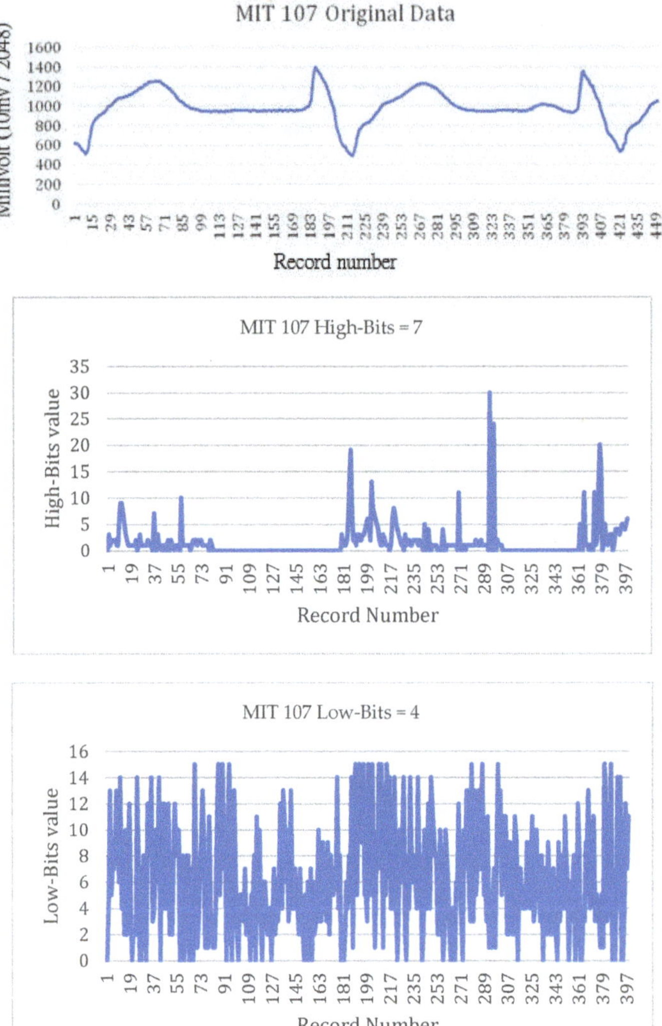

Figure 10. Worst case: MIT/BIH Arrhythmia Database recording 107 high bits = 10.

3.2. Block Size of Fixed-Block Mode

Whole high-bit data that are zero have the highest compression rate in this scheme; however, QRS segmentation causes the data to fail. To attain the optimum performance, the input data need an ECG data length of one heartbeat over time with QRS segmentation at the start of the data; this best-encoded high-bit data content employs QRS segmentation and an end tag only. This data length can be predicted because of changes in the heartbeat rate. Since using the QRS detection algorithm to segment the ECG signal makes this scheme complicated, this section addresses finding a prefixed- block size that has the most end tags to attain a higher encoded performance for each high-bit resolution with a 360 sampling rate in the MIT/BIH database.

High-bit data requires an efficient storage space to be encoded. Table 2 presents the results of different block sizes. Since the CR can be improved by increasing the block size, the block size properties for different CRs in Figure 11 suggest that diminishing marginal returns occur when the block size reaches 256 records. More storage space enhances the CR, but the effect diminishes.

Table 2. Block size and compression rate.

	100	101	102	103	104	105	106	107	108	109
128	2.197	2.143	2.152	2.094	1.916	1.809	1.912	1.596	1.913	1.819
256	2.214	2.164	2.170	2.111	1.931	1.813	1.923	1.591	1.925	1.822
384	2.218	2.169	2.174	2.116	1.934	1.813	1.922	1.588	1.925	1.822
512	2.220	2.172	2.177	2.119	1.936	1.814	1.922	1.586	1.923	1.824
640	2.221	2.174	2.179	2.122	1.937	1.814	1.922	1.586	1.923	1.826
768	2.221	2.174	2.179	2.122	1.937	1.813	1.921	1.584	1.922	1.826
896	2.222	2.175	2.180	2.123	1.938	1.814	1.921	1.584	1.923	1.827
1024	2.223	2.176	2.180	2.124	1.938	1.814	1.921	1.584	1.922	1.828
1152	2.223	2.177	2.181	2.124	1.939	1.813	1.921	1.583	1.922	1.829
1280	2.223	2.176	2.181	2.124	1.938	1.814	1.920	1.584	1.921	1.829
1408	2.223	2.177	2.181	2.125	1.938	1.814	1.920	1.583	1.922	1.829
1536	2.225	2.178	2.182	2.126	1.940	1.814	1.920	1.583	1.922	1.831
1664	2.224	2.178	2.182	2.126	1.940	1.813	1.920	1.583	1.922	1.830
1792	2.223	2.177	2.181	2.125	1.939	1.815	1.921	1.583	1.923	1.830
1920	2.224	2.178	2.182	2.126	1.940	1.814	1.920	1.582	1.922	1.831
2048	2.226	2.179	2.183	2.127	1.941	1.815	1.921	1.583	1.923	1.833

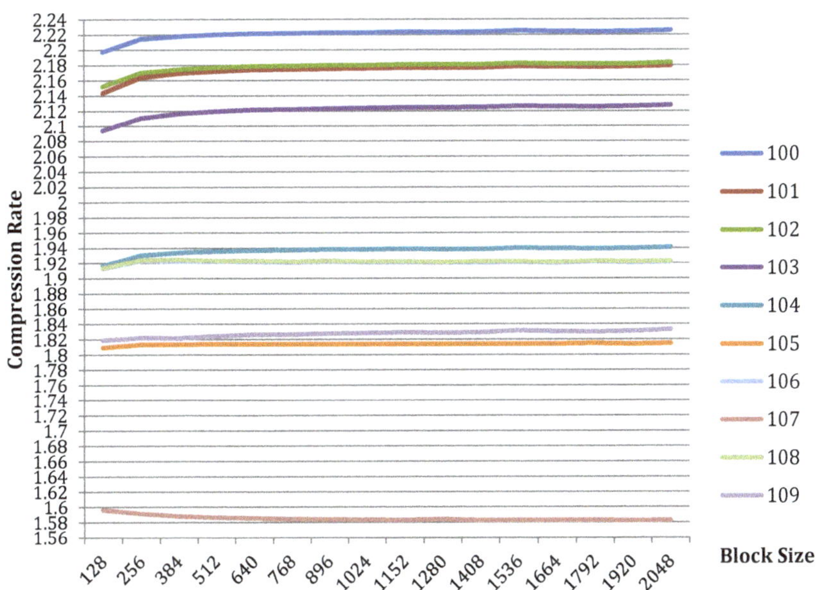

Figure 11. Compression rate trend.

3.3. Dynamic-Block Mode

Our experiments used high-bit resolution and block size, with the aim of eliminating as many records as possible using the end tag. MIT/BIH database recording 107 has the best CR with a block size of 128, whereas the other recordings have block sizes of 256. In dynamic-block mode, the block size is calculated by detecting the number of consecutive zeros produced by applying QRS to the high bits. Segmenting the high-bit streaming data with QRS complexes also provides useful information about the heart rate. The correct rate and CR for dynamic-block mode are shown in Table 3, which shows a positive correlation between the CR and heartbeat detection ability based on the numbers provided by the MIT database. However, while this feature works well for normal heartbeats, it performs poorly with MIT/BIH database recording 107, which features insignificant QRS complexes.

Table 3. Dynamic-block mode compression ratio (CR) and heartbeat detection rate.

HIGH-BITS		4 BITS		5 BITS		6 BITS		7 BITS	
	QRS	C.R	QRS	C.R	QRS	C.R	QRS	C.R	QRS
100	2273	1.52	2272	1.72	2272	2.00	2272	2.19	2272
101	1865	1.52	1863	1.74	1868	1.99	1868	2.13	1919
102	2187	1.53	2099	1.76	2148	1.98	2186	2.14	2187
103	2084	1.51	2084	1.72	2084	1.95	2083	2.07	2088
104	2229	1.53	2155	1.72	2231	1.88	2195	1.87	2132
105	2572	1.55	1351	1.67	2535	1.79	2492	1.72	2324
106	2027	1.52	1635	1.72	1895	1.92	1968	1.66	2049
107	2137	1.52	2132	1.59	2137	1.52	2105	1.25	1137
108	1774	1.56	503	1.77	1496	1.92	1906	1.67	1907
109	2532	1.54	1400	1.63	2528	1.81	2532	1.74	2410

4. Discussion

Lossy compression methods produce a much higher CR than lossless methods [26], with some methods achieving 10–20 times the compression with total errors under 10%. Table 4 shows the reported lossless compression results obtained from many different sources. Although the proposed method does not outperform other methods in terms of total compression, it can be easily implemented through ASIC or FPGA. In addition, although the heartbeat detection feature may not be entirely accurate, it may be suitable for some applications. The result could be used to trigger ECG data transmission, which could be followed by analysis at a medical center.

Existing PEM systems provide a feasible ECG recording solution that replaces wireless transmission with local storage. To determine its capabilities, fixed-block mode was implemented and verified on an Xilinx ML505 FPGA evaluation board (Figure 12). Arithmetic and logic units were used to implement the proposed scheme, which is shown as a block diagram in Figure 13. This implementation is triggered by data input from ADC conversion to reduce power consumption. An rs232 module was used as a communication interface between the desktop computer and evaluation board that transfers MIT/BIH arrhythmia data and verifies data compression.

Table 4 presents the results from various lossless ECG compression sources. Typical "entropy coding" is Huffman coding, which requires the Huffman tree to be written as part of the data. Differential pulse-code modulation (DPCM) uses a baseline of pulse-code modulation and a prediction function to encode data. DPCM linear prediction uses a discrete-time signal to estimate a linear function from previous samples. The entropy coding of second differences and orthogonal transforms—Compute Tomography (CT), Korhunen-Loeve transform (KLT), and Hibert transform (HT)—have the same time complexity, but orthogonal transforms is a frequency domain method; therefore, more computational power is used.

Table 4. Lossless compression rate method. DPCM: differential pulse-code modulation.

ECG Data Compression Schemes	Compression Rate
Entropy coding of 2nd differences	2.8
DPCM—delta coding with threshold	4
DPCM—LINEAR PREDICTION	2.5
Orthogonal transforms—CT, KLT, HT	3
Proposed encoding scheme	2.1

In Figure 13, the encoding scheme's delta coding module comprises CURRENT, PREVIOUS, and DELTA. The ROTATION and SEPARATION operation implement the zero separation module to generate the high and low-bit streams, which indicate to the other module that the input signal is ready. The low-bit stream is connected to the four-pin LowData_out interface, and can be stored on an external flash drive with four-bit writing. ZRLE uses MAX, HALF, ZERO, and ZERO-COUNTER to compress the high-bit stream from the previous module. ZERO-COUNTER is an increment that is used to count the number of consecutive zeros to trigger one MAX, HALF, or ZERO operation. The high-bit output stream can also be stored on an external flash drive using the storage module. Furthermore, Output_status_0 and Output_status_1 are used to identify three different output formats that appear in the high-bit stream pipeline system architecture: non-zero value status, maximal zero value status, and continual zero values with a non-zero value status. Two studies have developed lossless ECG data compression chips [26,35] and determined that a high operation clock is not needed (Table 5). Existing chips have low power consumption, and although this cannot currently be ascertained for the proposed chip, it is expected to be even lower.

Figure 12. Xilinx ML505 field-programmable gate array (FPGA) evaluation board.

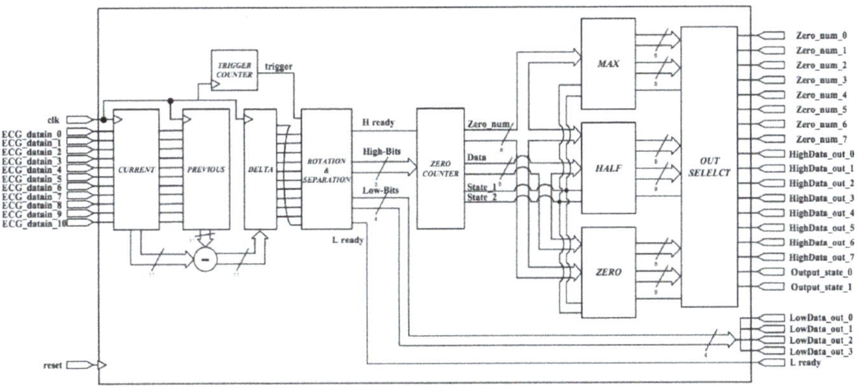

Figure 13. FPGA implementation.

Table 5. Compression operation clock.

	[35]	[36]	Proposed
Compress Rate	1.9	2.43	2.11
Operation Frequency	100M Hz	100M Hz	360 Hz
Gate counts (K)	13.4	3.57	1.16

5. Conclusions

Wearable ECG monitors do not provide sufficient computing power to detect critical P, QRS, and T events to allow a reliable cardiovascular assessment [37–42]. The proposed encoding uses a basic logic component, arithmetic and logic unit subtraction, rotation, comparator, and counter, and stores original real-time and lossless data for processing in clinical contexts. It is used to minimize the physical size and power consumption of a unit, and the encoding scheme offers superior efficiency to comparable software solutions. Since this scheme rotates only one bit, a simple circuit switch can be used to replace the rotating arithmetic logic unit (ALU). In addition, because each module is triggered by the previous one, the pipeline architecture achieves real-time ECG data encoding, and the scheme can be integrated into a wearable ECG monitor or PEM system. Replacing wireless modules with local storage in wearable ECG monitors (e.g., Holter device or PEM) reduces the overall footprint, power consumption, and battery weight. Existing PEM ASICs are suitable for implementing this design to minimize equipment size and power consumption while enhancing user convenience.

The designed scheme detects significant QRS complexes in a normal ECG signal. MIT records 100 and 103 have the best CR of approximately 2.1, whereas MIT 107 has the worst (i.e., 1.5), because it contains abnormal signals such as premature ventricular contractions and pauses. However, when high-bit resolution was split into five bits for MIT recording 107, both the CR and heartbeat detection correction increased. Despite its poor performance with abnormal ECG signals, the proposed approach can be used for Holter or PEM-based health care devices for occasional CVD assessment without data loss. FPGA was used to compress real ECG data from an ECG generator with a high sampling rate, and has reached a high CR (2.7 when a five-KHz sampling rate is used), as shown in Figure 14. This result confirms that an ADC-ready signal can be used as an operation clock when this scheme is implemented on a chip.

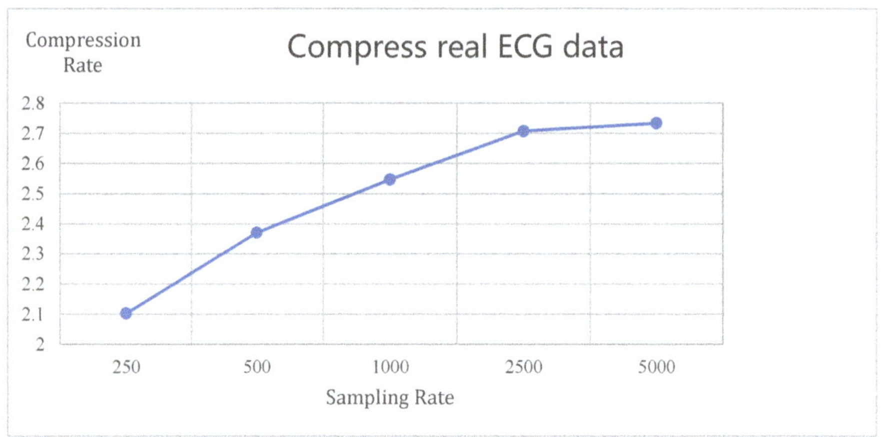

Figure 14. FPGA compresses real ECG data.

However, the actual performance did not meet expectations in dynamic-block mode. With an eight-bit high-bit resolution and a sampling rate of 360 Hz, the longest plate period of one heartbeat has 320 values for non-QRS segmentation, and it has the same storage requirements for the end tag (zero, zero) and zero encoding (zero and number). Dynamic-block mode could provide excellent performance with a higher data sampling rate or more high-bit resolution with a long continual zero values. MIT/BIH arrhythmia provides Physikalisch-Technische Bundesanstalt (PTB) databases with a one-kHz sampling rate and 16 bits of raw data resolution. However, this encoding scheme doesn't archive expected performance; the CR is only 1.98 because of eight high-bit resolutions and 8 eight high-bit resolutions. Figure 15 shows the MIT s0010_re probability of occurrence of each distinct symbol between the ECG (original) volume and delta volume variance. However, this record has high resolution, but the plate period needs eight bits in order to be present. Then, the maximal CR of s0010_re is restricted to the ratio of high bits and low bits. In fact, although MIT s0010_re has a 16-bit resolution, it still only uses nine bits to record data.

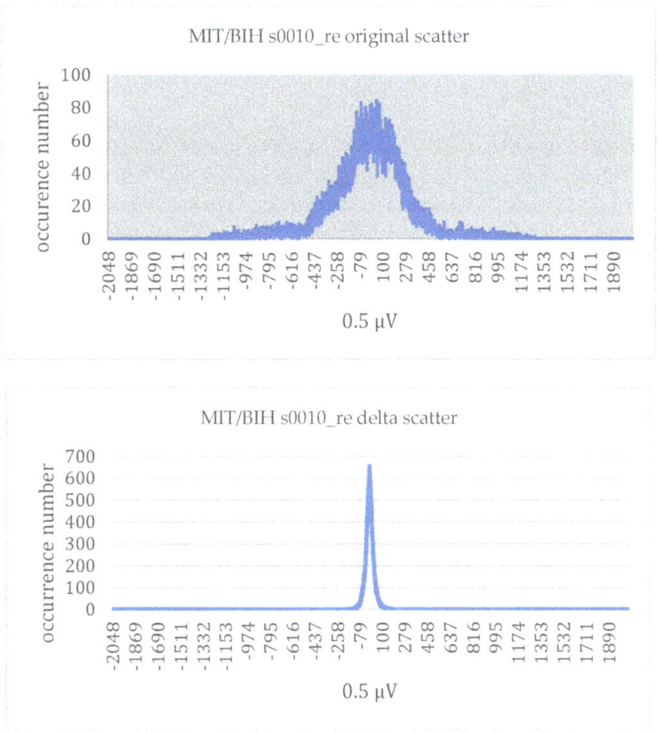

Figure 15. Comparison of probability of occurrence of each distinct symbol for ECG (original) volume and delta volume variance with MIT s0010_re.

The rough heartbeat rate detection feature of dynamic-block mode, which has a lower CR than that of fixed-block mode, could be used in medic alert applications. Furthermore, in order to enhance the CR, uncompressed low-bit data can be processed through a method such as Huffman coding. The optimal integration is to implement this scheme on an existing ADC chip or main processor, because it changes no input and output interfaces. Fixed-block mode can use a chip that is taped out from TSMC (Taiwan Semiconductor Manufacturing Company, Limited) to evaluate power consumption.

Author Contributions: H.W.F. conceived of the presented idea and developed the theory and performed the computations; C.C.L. verified the analytical methods and encouraged H.W.F. to investigate Lossless ECG data compression and supervised the findings of this work. All authors discussed the results and contributed to the final manuscript.

Funding: This research received no external funding.

Conflicts of Interest: The authors declare no conflicts of interest.

References

1. World Health Organization (WHO). *Global Status Report on Noncommunicable Diseases 2014*; World Health Organization: Geneva, Switzerland, 2014.
2. Dilaveris, P.E.; Gialafos, E.J.; Sideris, S.K.; Theopistou, A.M.; Andrikopoulos, G.K.; Kyriakidis, M.; Gialafos, J.E.; Toutouzas, P.K. Simple electrocardiographic markers for the prediction of paroxysmal idiopathic atrial fibrillation. *Am. Heart J.* **1998**, *135*, 733–738. [CrossRef]

3. Ren-Guey, L.; Yih-Chien, C.; Chun-Chieh, H.; Chwan-Lu, T. A mobile care system with alert mechanism. *IEEE Trans. Inf. Technol. Biomed.* **2007**, *11*, 507–517.

4. Rasid, M.F.A.; Woodward, B. Bluetooth telemedicine Processor for multichannel biomedical signal transmission via mobile cellular networks. *IEEE Trans. Inf. Technol. Biomed.* **2005**, *9*, 35–43. [CrossRef] [PubMed]

5. Wen, C.; Yeh, M.-F.; Chang, K.-C.; Lee, R.-G. Real-time ECG telemonitoring system design with mobile phone platform. *Measurement* **2008**, *41*, 463–470. [CrossRef]

6. Gradl, S.; Kugler, P.; Lohmuller, C.; Eskofier, B. Real-time ECG monitoring and arrhythmia detection using Android-based mobile devices. In Proceedings of the Annual International Conference of the IEEE Engineering in Medicine and Biology Society (EMBC), San Diego, CA, USA, 28 August–1 September 2012.

7. Scully, C.G.; Lee, J.; Meyer, J.; Gorbach, A.M.; Granquist-Fraser, D.; Mendelson, Y.; Chon, K.H. Physiological Parameter Monitoring from Optical Recordings with a Mobile Phone. *IEEE Trans. Biomed. Eng.* **2012**, *59*, 303–306. [CrossRef] [PubMed]

8. Mamaghanian, H.; Khaled, N.; Atienza, D.; Vandergheynst, P. Compressed sensing for real-time energy-efficient ECG compression on wireless body sensor nodes. *IEEE Trans. Biomed. Eng.* **2011**, *58*, 2456–2466. [CrossRef] [PubMed]

9. Oresko, J. Portable Heart Attack Warning System by Monitoring the St Segment via Smartphone Electrocardiogram Processing. Ph.D. Thesis, University of Pittsburgh, Pittsburgh, PA, USA, 2010.

10. Yoo, J.; Yan, L.; Lee, S.; Kim, H.; Yoo, H.-J. A wearable ECG acquisition system with compact planar-fashionable circuit board-based shirt. *IEEE Trans. Inf. Technol. Biomed.* **2009**, *13*, 897–902. [CrossRef] [PubMed]

11. Rita Paradiso, G.L.; Taccini, N. A Wearable Health Care System Based on Knitted Integrated Sensors. *IEEE Trans. Inf. Technol. Biomed.* **2005**, *9*, 337–344. [CrossRef] [PubMed]

12. Abo-Zahhad, M. ECG Signal Compression Using Discrete Wavelet Transform. In *Discrete Wavelet Transforms—Theory and Applications*; InTech: London, UK, 2011; Chapter 8; pp. 143–168, ISBN 978-953-307-185-5.

13. Stearns, S.D.; Tan, L.-Z.; Magotra, N. Lossless compression of waveform data for efficient storage and transmission. *IEEE Trans. Geosci. Remote Sens.* **1993**, *31*, 645–654. [CrossRef]

14. Logeswaran, R.; Eswaran, C. Performance survey of several lossless compression algorithms for telemetry applications. *Int. J. Comput. Appl.* **2001**, *23*, 1–9. [CrossRef]

15. Sayood, K. *Introduction to Data Compression*, 3rd ed.; Morgan Kaufmann Publishers: San Francisco, CA, USA, 2006.

16. Aydin, M.C.; Cetin, A.E.; Koymen, H. ECG data compression by subband coding. *Electron. Lett.* **1991**, *27*, 359–360. [CrossRef]

17. Tai, S.C. Six band subband coder on ECG waveforms. *Med. Biol. Eng. Comput.* **1992**, *30*, 187–192. [CrossRef] [PubMed]

18. Rajoub, B.A. An efficient coding algorithm for the compression of ECG signals using the wavelet transform. *IEEE Trans. Biomed. Eng.* **2002**, *49*, 355–362. [CrossRef] [PubMed]

19. Velasco, M.; Roldn, F.; Llorente, J.; Barner, K.E. Wavelet Packets Feasibility Study for the Design of an ECG Compressor. *IEEE Trans. Biomed. Eng.* **2007**, *54*, 766–769. [CrossRef] [PubMed]

20. Benzid, R.; Marir, F.; Boussaad, A.; Benyoucef, M.; Arar, D. Fixed percentage of wavelet coefficients to be zeroed for ECG compression. *Electron. Lett.* **2003**, *39*, 830–831. [CrossRef]

21. Pooyan, M.; Taheri, A.; Goudarzi, M.; Saboori, I. Wavelet compression of ECG signals using SPIHT algorithm. *Int. J. Signal Process.* **2005**, *1*, 219–225.

22. Miaou, S.-G.; Yen, H.-L.; Lin, C.-L. Wavelet-based ECG compression using dynamic vector quantization with tree codevectors in single codebook. *IEEE Trans. Biomed. Eng.* **2002**, *49*, 671–680. [CrossRef] [PubMed]

23. Dakua, S.P.; Sahambi, J.S. Lossless ECG Compression for Event Recorder Based on Burrows-Wheeler Transformation and Move-To-Front Coder. *Int. J. Recent Trends Eng.* **2009**, *1*, 120.

24. Kannan, R.; Eswaran, C. Lossless compression schemes for ECG signals using neural network predictors. *EURASIP J. Appl. Signal Process.* **2007**, *2007*, 102. [CrossRef]

25. Arnavut, Z. Lossless and Near-Lossless Compression of ECG Signals with Block-Sorting Techniques. *Int. J. High Perform. Comput. Appl.* **2007**, *21*, 50–58. [CrossRef]

26. Koski, A. Lossless ECG encoding. *Comput. Methods Programs Biomed.* **1997**, *52*, 23–33. [CrossRef]

27. Giurcăneanu, C.D.; Tăbuş, I.; Mereuţă, Ş. Using contexts and R-R interval estimation in lossless ECG compression. *Comput. Methods Programs Biomed.* **2002**, *67*, 177–186. [CrossRef]
28. Miaou, S.-G.; Chao, S.N. Wavelet-Based Lossy-To-Lossless ECG Compression in a Unified Vector Quantization Framework. *IEEE Trans. Biomed. Eng.* **2005**, *52*, 539–543. [CrossRef] [PubMed]
29. MIT. MIT-BIH Arrhythmia Database. Available online: http://www.physionet.org/physiobank/database/mitdb/ (accessed on 19 December 2011).
30. Cox, J.R.; Nolle, F.M.; Fozzard, H.A.; Oliver, G.C., Jr. AZTEC, a Preprocessing Program for Real-Time ECG Rhythm Analysis. *IEEE Trans. Biomed. Eng.* **1968**, *2*, 128–129. [CrossRef]
31. Hamming, R.W. *Coding and Information Theory*; Prentice-Hall: Upper Saddle River, NJ, USA, 1980.
32. Chang, G.C.; Lin, Y.D. An Efficient Lossless ECG Compression Method Using Delta Coding and Optimal Selective Huffman Coding. In Proceedings of the 6th World Congress of Biomechanics (WCB), Singapore, 1–6 August 2010; pp. 1327–1330.
33. Kavousianos, X.; Kalligeros, E.; Nikolos, D. Optimal Selective Huffman Coding for Test-Data Compression. *IEEE Trans. Comput.* **2007**, *56*, 1146–1152. [CrossRef]
34. Bradley, S.D. Optimizing a scheme for run length encoding. *Proc. IEEE* **1969**, *57*, 108–109. [CrossRef]
35. Chioukh, L.; Boutayeb, H.; Deslandes, D.; Wu, K. Noise and Sensitivity of Harmonic Radar Architecture for Remote Sensing and Detection of Vital Signs. *IEEE Trans. Microw. Theory Tech.* **2014**, *62*, 1847–1854. [CrossRef]
36. Chen, S.L.; Lee, H.Y.; Chen, C.A.; Huang, H.Y.; Luo, C.H. Wireless Body Sensor Network with Adaptive Low-Power Design for Biometrics and Healthcare Applications. *IEEE Syst. J.* **2009**, *3*, 398–409. [CrossRef]
37. Chen, S.L.; Wang, J.G. VLSI implementation of low-power cost-efficient lossless ECG encoder design for wireless healthcare monitoring application. *Electron. Lett.* **2013**, *49*, 91–93. [CrossRef]
38. Thong, T.; McNames, J.; Aboy, M.; Goldstein, B. Prediction of paroxysmal atrial fibrillation by analysis of atrial premature complexes. *IEEE Trans. Biomed. Eng.* **2004**, *51*, 561–569. [CrossRef] [PubMed]
39. Tsipouras, M.G.; Fotiadis, D.I.; Sideris, D. Arrhythmia classification using the RR-interval duration signal. In Proceedings of the IEEE Computers in Cardiology, Memphis, TN, USA, 22–25 September 2002; pp. 485–488.
40. Bashour, C.; Visinescu, M.; Gopakumaran, B.; Wazni, O.; Carangio, F.; Yared, J.P.; Starr, N. Characterization of premature atrial contraction activity prior to the onset of postoperative atrial fibrillation in cardiac surgery patients. *Chest* **2004**, *126*, 831S–832S. [CrossRef]
41. De Chazal, P.; Dwyer, M.O.; Reilly, R.B. Automatic classification of heartbeats using ECG morphology and heartbeat interval features. *IEEE Trans. Biomed. Eng.* **2004**, *51*, 1196–1206. [CrossRef] [PubMed]
42. Krasteva, V.T.; Jekova, I.I.; Christov, I.I. Automatic detection of premature atrial contractions in the electrocardiogram. *Electrotech. Electron.* **2006**, *9–10*, 49–55.

Article

Foot-Mounted Inertial Measurement Units-Based Device for Ankle Rehabilitation

Alfonso Gómez-Espinosa *, Nancy Espinosa-Castillo and Benjamín Valdés-Aguirre

Tecnologico de Monterrey, Campus Querétaro, Ave. Epigmenio González 500, Fracc. San Pablo, Querétaro, QRO 76130, Mexico; A01206782@itesm.mx (N.E.-C.); bvaldesa@itesm.mx (B.V.-A.)
* Correspondence: agomeze@itesm.mx; Tel.: +52-442-238-3302

Received: 6 October 2018; Accepted: 19 October 2018; Published: 24 October 2018

Abstract: Ankle sprains are frequent injuries that occur among people of all ages. Ankle sprains constitute approximately 15% of all sports injuries, and are the most common traumatic emergencies. Without proper treatment and rehabilitation, a more severe sprain can weaken the ankle, making it more likely for new injures, and leading to long-term problems. In this work, we present an inertial measurement units (IMU)-based physical interface for measuring the foot attitude, and a graphical user interface that acts as a visual guide for patient rehabilitation. A foot-mounted physical interface for ankle rehabilitation was developed. The physical interface is connected to the computer by a Bluetooth link, and provides feedback to the patient while performing dorsiflexion, plantarflexion, eversion, and inversion exercises. The system allows for in-home rehabilitation at an affordable price while engaging the patient through active therapy. According to the results, more consistent rehabilitation could be achieved by providing feedback on foot angular position during therapy procedures.

Keywords: IMU; inertial sensor; gyroscope; accelerometer; rehabilitation; ankle sprain; wearable electronic device; Unity; Arduino; telerehabilitation

1. Introduction

Ankles sprains are common injuries that can produce temporary or permanent impairments. An ankle sprain occurs when the ligaments are torn because of being stretched beyond their limits. Ankle sprain causes approximately 15 to 20% of sports injuries, when considering all the different grades of tears [1]. It can take several weeks or even months for a sprained ankle to heal. When a sprain is not properly treated, a more severe sprain can occur and permanently weaken the ankle, making it more likely for new injures to take place. Due to the lack of correct treatment, repeated ankle sprains can lead to long-term problems, such as chronic ankle pain, ongoing instability, and arthritis [1]. In this case, rehabilitation is the part of the treatment that helps restore a patient to the regular use of their injured extremity through eliminating the impairments caused by the injury, when there has been no permanent damage. The problem is that unsupervised rehabilitation is prone to mistakes. Well-performed medical treatment can be undermined by poor rehabilitation [2].

Ankle movement can be described as a three-dimensional object with three degrees of freedom. The basic movements used in ankle rehabilitation are: abduction, adduction, plantarflexion, dorsiflexion, inversion, and eversion.

The term sprain, also known as twisting or entorsis, indicates ligament injury by traction, which is executed by an indirect mechanism that extremely stretches the ligament structure beyond what the ligaments can tolerate. The mechanism of this injury is related to the inversion or eversion movements. The inversion movement is the most common acute injury that damages the lateral ligaments of the ankle. Its inadequate treatment can lead to problems, such as a decreased range of motion, joint instability, and pain [1].

Depending on the degree of the injury, this traumatism can be catalogued into the following categories [1]:

- Grade I: Involves the stretching of any ligament without tearing and slight signs of pain and/or inflammation.
- Grade II: Involves the partial tearing of one or more ligaments and moderate pain and inflammatory signs.
- Grade III: Involves full tear of ligaments and joint instability; pain and inflammatory signs are significant, and there is a loss of ankle function and mobility.

The diagnosis of ankle sprain starts with the evaluation of the situation and mechanism of the injury. This is done to determine the current degrees of possible ankle movement, and the required treatment for this specific situation. This evaluation is based primarily on physical exploration using techniques such as inspection, palpation, mobilization, and radiology. Due to the frequency of this injury, it is necessary to develop a set of strengthening exercises for the rehabilitation of the ankle that allows recovering the amplitude of the movement of the joint and helps strengthen the muscles of this extremity.

Telerehabilitation is the clinical application of Information and Communication Technologies (ICT) to provide rehabilitation services remotely and without the presence of a therapist all of the time. Telerehabilitation has been proven to be technically feasible; the case has been made as to how this technology can improve access to health services, as well as to extend the reach of clinicians beyond the physical walls of a traditional health care facility [1].

Devices that are used for ankle sprain rehabilitation can be grouped by complexity. Examples of low-complexity devices include elastic bands, roller foams, and wobble boards. Intermediate complexity devices provide more consistent results; in these devices, the patient plays a passive role in the rehabilitation process. Examples of intermediate devices include the DOF JACE Ankle A330 CPM system and the DOF Optiflex Ankle CPM system. High-complexity devices have high capabilities for data acquisition, storage, transmission, and providing written feedback. However, some disadvantages include being expensive and bulky, which restrict the use outside of a clinic. An example of these systems is the Biodex Multi-Joint System 4 Pro [1]. Besides the lack of portability, perhaps the most important disadvantage is that intermediate and complex systems work in a continuous passive motion basis [1,3,4]. This reduces the type of therapy where they can be used. Portable devices are needed for the active rehabilitation of ankle injuries; these devices could reduce the cost of healthcare for patients [5].

Inertial measurement units (IMU) can be used to make portable rehabilitation devices. They have been extensively used in position and attitude estimation, allowing for pedestrian navigation tracking [6–13], gait analysis [14–20], foot pose estimation [21], foot clearance estimation [22], wearable devices for Game Play application [23], and detecting foot strike in recreational runners [24]. A problem of inertial measurement units is that they present biases and other systematic errors that are responsible for position and attitude estimation errors. To provide robust distortion-free and refined absolute position and orientation vectors, data fusion is used to combine the measurements from the three-axis gyroscope, three-axis geomagnetic sensor, and three-axis accelerometer enclosed in the IMU [17]. Approaches used with data fusion include Kalman filters [6,9,15], clustering algorithms [7], and hidden Markov models [8]. Some of these approaches rely on additional information, such as map information [8,10], multiple IMU [11], biomechanical models [25], or light detection and ranging (LIDAR) systems [12]. A major challenge of micro-electro-mechanical systems (MEMS) position estimation concerns how to restrain the rapid accumulation of navigation errors. Without an error-resetting algorithm, this error rapidly increases. For pedestrian navigation tracking and gait analysis, a zero velocity update (ZVU) algorithm could be employed as an effective way to correctly detect the stance phase of each gait cycle and compensate for speed bias errors [6,9]. Another source for errors in IMU sensors comes from magnetic disturbances that might take place in indoor environments such as medical facilities, which becomes a problem for IMU clinical applications. Two approaches that are used to reduce the impact of magnetic disturbance are: magnetic anomaly detection algorithms [13], and a method based on analytical solutions in combination with parameterization for corrections [26]; both are used as part of data fusion.

IMU use related to rehabilitation therapy can be separated into systems that provide explicit biofeedback for therapy and those that aim to help with the diagnosis and detection of pathologies. In the first group, we can find systems for treating foot drop syndrome [27], hip and knee rehabilitation exercises [28], improving therapy adherence [29], improving gait in patients with walking disabilities [18], exoskeletons [30], and even therapies for animals [20]. In the second group, we can find systems that help diagnose gait pathologies [16], explore elderly gait [19], and assess human gait parameters [14]. Most of the above used foot-mounted devices or insoles. Chirakanphaisarn, Seel, Long [18,27,30] created systems that depend on other sensors besides the IMU. Giggins [28] tested their system with a single IMU mounted on the foot and another IMU incorporated in knee and hips, and Shepherd [29] used a single IMU-based approach as well. The case has been made that for some rehabilitation, a single IMU could suffice for posture and motion recognition [28]. Of relevance is Seel [27], which explored foot movement for rehabilitation, and used IMU to map foot pitch and roll to dorsiflexion, eversion, and inversion movements, which are also used in rehabilitation exercises for ankle sprains.

In this paper, we developed a foot-mounted inertial sensor device to determine the foot attitude with the aim of making an active rehabilitation session following the same process that a therapist would. We develop a graphical software interface to provide feedback on common rehabilitation movements that can be described with two degrees of freedom: dorsiflexion, plantarflexion, inversion, and eversion. The work has been done for the active free stage of rehabilitation, which is the stage where the patient makes the movements without any opposing force [1].

For the remaining sections of this paper, in Section 2, the materials, methods and main concepts are presented. Section 3 presents the foot attitude biofeedback device description. Section 4 shows and describes the experimental results. Finally, in Section 5, concluding remarks are provided.

2. Materials and Methods

2.1. Inertial Sensor

2.1.1. Inertial Measurement Units

Inertial measurement units (IMUs) have been widely used in mobile robot navigation [31–33], and can be used to make portable rehabilitation devices. IMU devices measure acceleration, angular rates, and magnetic field vectors in their own three-dimensional local coordinate system, and, with proper calibration, represent an orthonormal base that is aligned with the outer sensor casing [34]. Some commercial devices provide the orientation estimation of this sensor for a global fixed coordinate system. They do this by using algorithms that provide orientation information in the form of quaternion, rotation matrix, or Euler angles. These algorithms employ a strap-down integration of the angular rates to obtain a first estimate of the orientation. Also, the angular position drifts in the inclination part of the IMU's orientation, which is eliminated using the assumption that the measured acceleration is dominated by gravitational acceleration (i.e., for low-acceleration ranges) [34].

Ankle movement can be described as a three-dimensional object with three degrees of freedom. The basic movements used in ankle rehabilitation (abduction, adduction, plantarflexion, dorsiflexion, inversion, and eversion) are shown on the foot anatomical reference frame in Figure 1. This work is focused on ankle movements that can be described with two degrees of freedom: dorsiflexion, plantarflexion, inversion, and eversion. Dorsiflexion and plantarflexion correspond to the angular movement denominated as pitch, which is performed around the "Y" axis, with normal ranges of $20°$ and $50°$, respectively. Inversion and eversion relate to the angular movement denominated as roll, which is performed around the "X" axis, with normal ranges of $45°$ and $20°$, respectively. For this work, the main interest lies in the active free stage of rehabilitation, which is the stage where the patient makes the movements without any opposing force [1].

A fundamental problem in IMU-based human motion analysis is that the IMU's local coordinate axes are not aligned with any physiologically meaningful axis. However, a common approach is to do this via calibration postures and/or calibration movements. Some authors make the patient stand with vertical, straight legs for a few seconds, and use the acceleration measured during that time interval to determine the local coordinates of the segment's longitudinal axis [34].

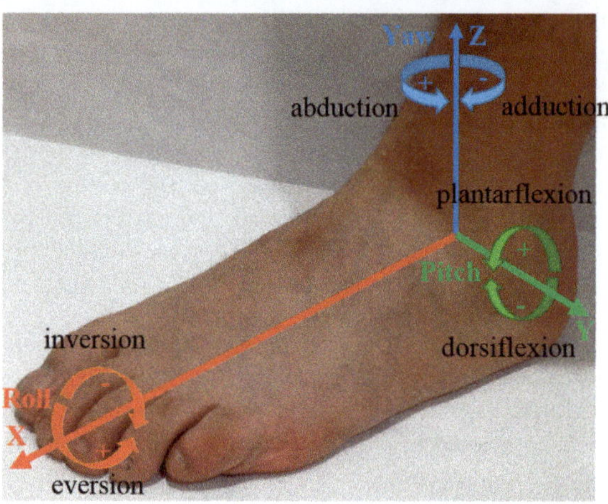

Figure 1. Foot anatomical reference frame, and the six basic movements: abduction, adduction, plantarflexion, dorsiflexion, inversion, and eversion. Sign of the rotation is specified according to the mathematical definition of the counterclockwise (positive) or clockwise (negative) rotation.

2.1.2. Estimation of the Inertial Measurement Variable

An MPU-6050 device was chosen for developing the rehabilitation device due to its commercial availability in local stores and low price. The MPU-6050 consists of six degrees of freedom (6DOF), which combine a three-axis accelerometer and a three-axis gyroscope. The main characteristics of this component are 16-bit digital analog converters (ADC), and that the component can communicate by both SPI and I2C bus, which makes it easy to access the measured data. On the other hand, the accelerometer range can be set to ±2 g, ±4 g, ±8 g, and ±16 g, and the gyroscope range can be set to $\pm250°/s$, $\pm500°/s$, $\pm1000°/s$, and $\pm2000°/s$; for the best resolution, ranges of ±2 g and $\pm250°/s$ were selected.

To use the device, the angular velocity and angle with respect to the horizontal position must be obtained from the IMU. The angular velocity $\omega = (\omega x, \omega y, \omega z)$, is directly provided by the three-axis gyroscope in the IMU, but the angular position $\theta = (\beta, \alpha, \gamma)$ needs to be calculated through the process explained below.

Following the methodology presented in Seel [35], defining pitch angle α and roll angle β, the IMU reference frame of the foot anatomical reference coordinates x_f, y_f has to be determined. The IMU was fixed to the foot to assure that the IMU sensor axis x_s lies in the sagittal plane of the foot, and then we proceeded to determine the x_f, y_f axis as follows: at the foot flat position, the IMU accelerometer readings $a(t)$ are integrated over time, and the resulting vector is normalized to the unit magnitude z_{ff}, applying the Euclidian norm:

$$\hat{z}_{ff} = \sum_{t \in [foot\ flat]} a(t) \tag{1}$$

$$z_{ff} = \hat{z}_{ff} / \|\hat{z}_{ff}\|^2 \tag{2}$$

During the calibration procedure, at the foot flat position, the foot is at rest, and z_{ff} is nearly vertical. Then, y_f is calculated as follows:

$$\hat{y}_f = z_{ff} \times x_s, \tag{3}$$

$$y_f = +\hat{y}_f / \|\hat{y}_f\|^2 \text{ for a left foot,} \tag{4}$$

$$y_f = -\hat{y}_f / \|\hat{y}_f\|^2 \text{ for a right foot} \tag{5}$$

Similarly, x_f is calculated as follows:

$$\hat{x}_f = z_{ff} \times x_s \times z_{ff}, \tag{6}$$

$$x_f = \hat{x}_f / \|\hat{x}_f\|^2 \tag{7}$$

For each foot exercise, a strap-down integration of the angular velocity ω is started that produces the rotation matrix $R_{ff}(t)$. By transforming x_f to the reference frame, we calculate the pitch angle α:

$$\alpha(t) = \frac{\pi}{2} - \sphericalangle\left(z_{ff}, R_{ff}(t)x_f\right), \tag{8}$$

$$= \arcsin(z_{ff}^T R_{ff}(t)x_f), \in \left[-\frac{\pi}{2}, +\frac{\pi}{2}\right] \tag{9}$$

Similarly, we calculate the roll angle β:

$$\beta(t) = \frac{\pi}{2} - \sphericalangle\left(z_{ff}, R_{ff}(t)y_f\right), \tag{10}$$

$$= \arcsin(z_{ff}^T R_{ff}(t)y_f), \in \left[-\frac{\pi}{2}, +\frac{\pi}{2}\right] \tag{11}$$

The IMU provides two information sources that allow the estimation of the angular position θ of the IMU local coordinate system. The first is through the three-axis accelerometer signals $\mathbf{g} = (gx, gy, gz)$, which measure the vector of the acceleration of gravity, and trigonometric relations are used to calculate one estimation for the angular position θ. The second is through the three-axis gyroscope, which measures the angular velocity ω, whose numerical integral with respect to time gives an approximation of the angular rotation since the beginning of the integration, and is used to compute a second estimation for the angular position θ. These means that in the first source, information is highly noisy due to the accelerations caused by vibrations that can occur in rehabilitation procedures. This makes it unreliable in short periods, but this same source is reliable on average over a longer period of time. The second source of information is very reliable in short periods, but accumulates offset errors as time progresses. For this reason, the MPU-6050 incorporates a digital motion processor (DMP), which is an internal processor that executes MotionFusion algorithms to combine the accelerometer and gyroscope data together to minimize the effects of the errors that are inherent in each sensor, avoiding having to compute the filters by an external processor. The DMP computes the results in terms of quaternions; it can also convert the results to Euler angles, and perform other filtering computations with the data as well [36].

The DMP eliminates the need to perform computationally expensive calculations on the Arduino side. The problem is that there is little information available on the inner workings of the DMP. This problem was addressed by using a library created by Jeff Rowberg [37]. The library can calculate the sensor's orientation with respect to a global fixed coordinate system in Euler angles from the raw data numerical values from the DMP as given by the sensors, without any additional type of filter or processing. Jeff Rowberg's library also allows specifying the sensor sensitivity and measurement ranges for each variable.

Some IMU consist of a tri-axial accelerometer, a tri-axial gyro, and a tri-axial magnetic sensor. The initial conditions for gyro integration can be given by the accelerometer, which is used in combination with the magnetic sensor. However, ferromagnetic materials near the IMU critically disturb the magnetic sensor output [26,38]. We use only gyroscopes and accelerometers, and therefore do not rely on a homogeneous magnetic field, and the complete orientation of the IMU with respect to a global reference coordinate system is calculated using a fusion algorithm that combines gyroscope and accelerometer measurements.

It has been demonstrated that: "in a real-time assessment of foot orientation by accelerometers and gyroscopes, validated with respect to a conventional optical motion capture system in trials with subjects walking, at different velocities, root mean square deviations of less than $4°$ were found in all scenarios, while values near $2°$ were found in slow walking" [35]. Taking this into consideration, regular therapy procedures can be correctly performed considering this angular precision for the exercises.

3. Foot Attitude Biofeedback Device

The solution proposed is a physical interface consisting of an electric circuit to measure the degrees of ankle movement and an application that provides continuous graphical feedback to the patient during his rehabilitation exercises. The physical interface provides feedback on dorsiflexion, plantarflexion, eversion, and inversion exercises.

The software components that were used to develop the app and communicate with the hardware were the Arduino Software (IDE) version 1.6.9 (Open-source general public license), the game development platform Unity 5.5 (Unity Technologies, 30 3rd Street, San Francisco, CA 94103, United States), and SQLite3 (Hwaci, 6200 Maple Cove Ln, Charlotte, NC 28269, EE. UU.), which is a relational database management system that is compatible with the properties ACID.

The hardware components that were used in the device were the USB-based microcontroller development system Teensy 2.0, the MPU-6050 sensor, the Bluetooth HC-05, three capacitors of 330 mF, 100 nF, and 10 mF, and two voltage regulators, 7805CT and LM2936.

3.1. MPU-6050

The MPU-6050 sensor contains micro-electro-mechanical systems (MEMS), a 16-bit resolution three-axis accelerometer, and a 16-bit resolution three-axis gyroscope in a single chip. They capture the x, y, and z channel at the same time, which means that it is an IMU of six degrees of freedom (6DOF). The operational ranges for the accelerometers and gyroscope were respectively selected to ± 2 g, and ± 250 °/s. Consequently, accelerometer least significant bit sensitivity was 8192 LSB/mg, and gyroscope least significant bit sensitivity was 131 LSB/(°/s). The sampling rate was selected to 100 Hz.

The MPU-6050 works with 3.3 volts; it contains 16-bits analog to digital conversion hardware for each channel, and it uses the I2C bus to interface with the Teensy 2.0 and Arduino IDE. The addresses that were used in I2C are shown in Table 1.

Table 1. Addresses for I2C.

Pin AD0	Address 12C
AD0 = HIGH (5 V)	0×69
AD0 = LOW (GND or NC)	0×68

The ADDR pin internally in the module has a resistance to GND, so its connection was not necessary, because the default address is 0×68.

The raw values of the accelerometer and gyroscope are read when the sleep mode is disabled. This sensor contains a 1024 byte FIFO buffer to the sensor values, which can be programmed to be placed in the FIFO buffer, and then they can be read by Arduino.

The FIFO buffer is used together with the interrupt signal. If the MPU-6050 places data in the FIFO buffer, it signals the Arduino with the interrupt signal, so the Arduino knows that there is data in the FIFO buffer waiting to be read [39].

3.2. Libraries for the MPU-6050

To know the angular movements, it was necessary to work with the library developed by Jeff Rowberg; this library works with an additional library for I2C communication called I2Cdev. The libraries are used to read the accelerometer/gyro data and handle all of the calculations to convert the raw values of the angular velocity and acceleration into angles for determining the foot attitude. Correct placement of the IMU device is assured by fastening the MPU-6050 to the instep according to the orientation presented in Figure 2, considering the foot anatomical reference frame shown in Figure 1.

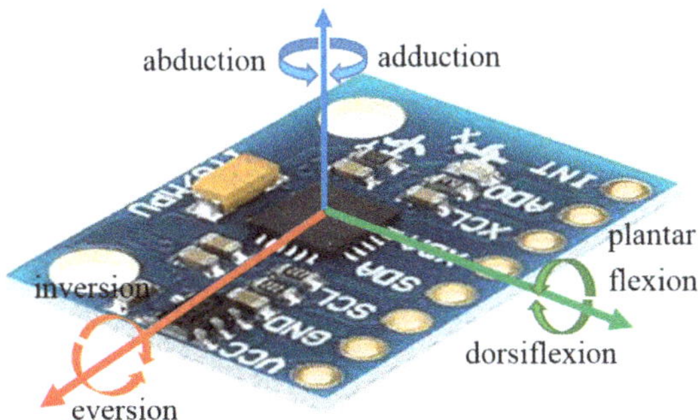

Figure 2. Correct placement of the IMU device MPU-6050, considering the approximate foot anatomical reference frame.

3.3. Calibration

The MPU-6050 needs to be calibrated before it can be used properly. The goal is to remove the zero error; this is where the sensor is leveled, but it detects that it is slightly angled. Therefore, we need to adjust the offsets in order to compensate for this error. We used a script made by Luis Rodenas [40] that calibrates the MPU-6050 and solves this issue. To use the program, the script needs to be uploaded to the accel–gyro module, and the module should be placed in a flat and level position. The program will make an average of a few hundred readings and display the offsets that are required to remove zero error. This calibration is carried out before every trial, while an IMU is attached to the foot, and the foot must be completely still to avoid introducing measurement errors. It is essential to properly place the sensor to ensure accurate measurements.

3.4. Connections

The MPU-6050 and the Teensy 2.0 were connected (Table 2), and the Bluetooth and the Teensy 2.0 module were connected as well (Table 3). The connections between the hardware components are shown Figure 3. The prototype consisted of a band (MPU-6050) that was connected to a small case that contained the rest of the components (Figure 4).

Table 2. Connections between the MPU-6050 and Teensy 2.0 modules.

MPU-6050	Teensy 2.0
VCC	3.3 V
GND	GND
SCL	D0
SDA	D1

Table 3. Connection between the Bluetooth and Teensy 2.0 modules.

Bluetooth HC-05	Teensy 2.0
VCC	5 V
GND	GND
RXD	B1
TXD	D2

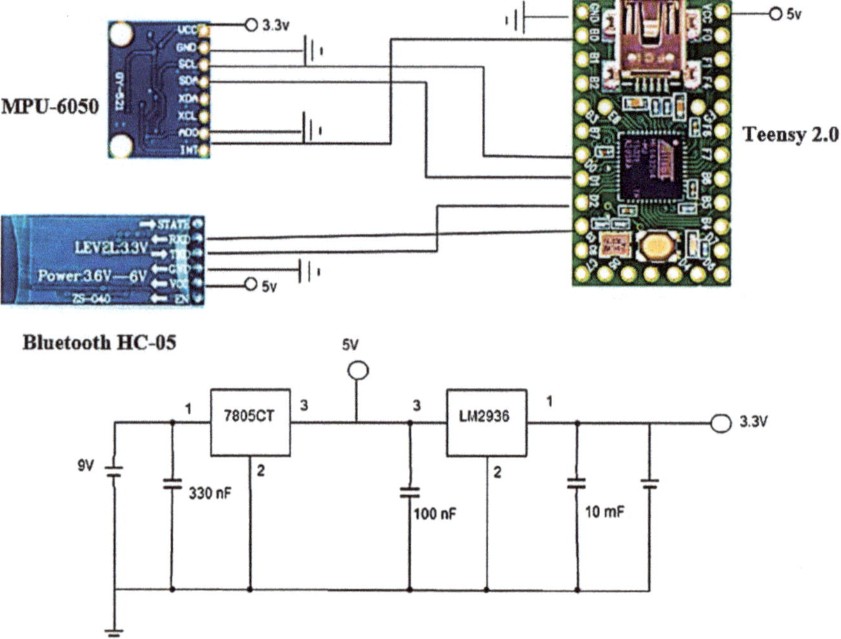

Figure 3. Connections diagram.

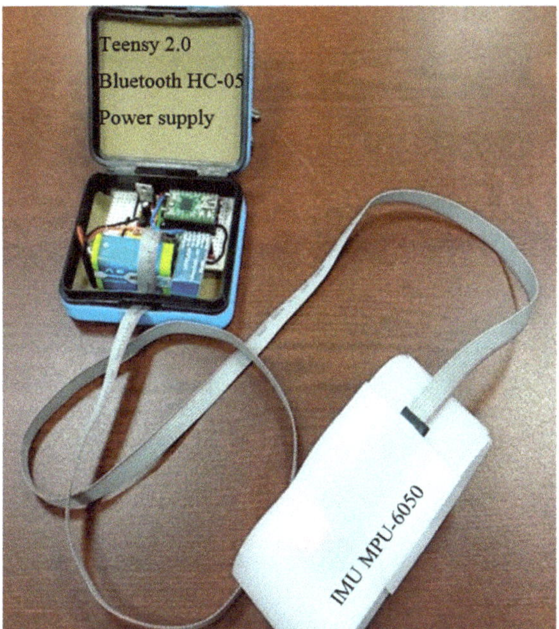

Figure 4. Portable foot attitude measuring device.

3.5. Communication between Arduino and Unity

Unity 5.5 is used to process the data from Arduino and generate a graphic user interphase (GUI), which is used for calibration and during the exercises (Figure 5).

To integrate Unity 5.5 with the Arduino IDE, it is necessary to use the serial port with the class SerialPort that mediates the communication in C#. Unity doesn't include the necessary libraries by default. Therefore, one must include the full NET 2.0 library in Unity. To establish the communication between Unity and the Arduino, Steven Cogswell's ArduinoSerialCommand library was used.

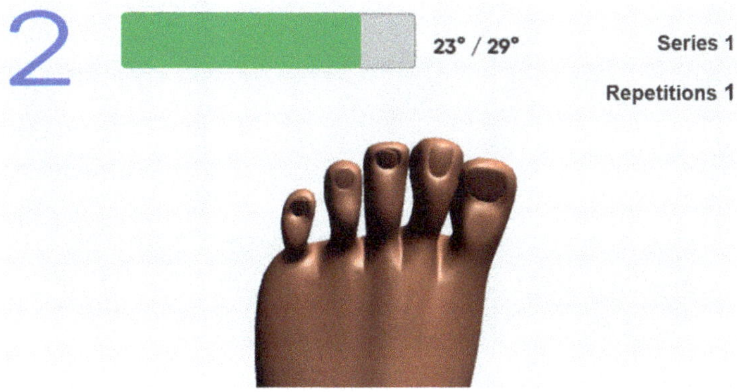

Figure 5. Graphical user interface.

3.6. SQLite

SQLite is a library that implements a self-contained, serverless, zero-configuration, transactional SQL database engine. The database is defined by the entity relationship model that is shown in Figure 6.

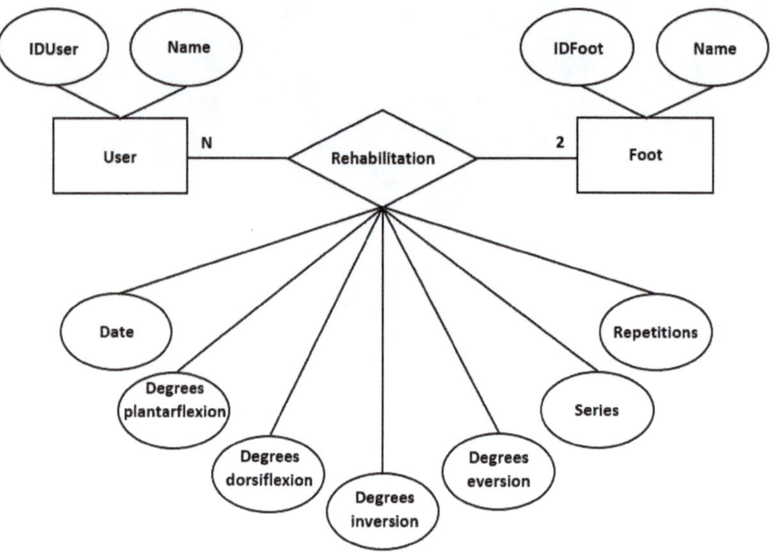

Figure 6. Entity relationship model.

The final product that was used for the evaluation sessions can be seen in Figure 7.

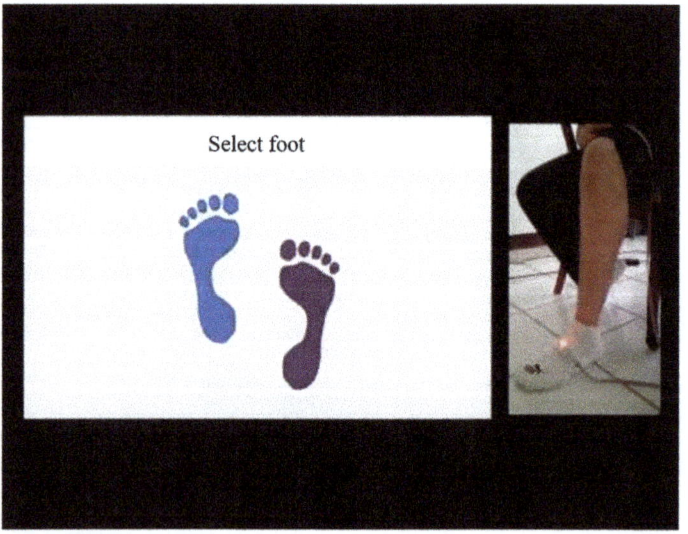

Figure 7. Experimental validation showing the graphical user interface and the patient.

4. Experimental Results and Discussion

4.1. Motion Range Evaluation

To evaluate whether the physical interface can effectively detect the movement required for rehabilitation, separate rehabilitation and/or strengthening sessions were conducted with a group of 10 people age 34 ± 16, gender (five male, five female), and a range of ankle motion depending on previous ankle sprain events. Six were under rehabilitation, and four had old ankle sprains that had not been treated by professional therapists. The research was approved by the University's Evaluation Committee on 2 February 2017 under the project identification code: A01206782-2017EM.

Each patient performed a total of four sessions of rehabilitation and/or strengthening, with two different types of assessment for each foot: the assessment with a therapist and the evaluation with the application. Each session consisted of basic 2DOF movements, which were identified as dorsiflexion, plantarflexion, inversion, and eversion. Both types of evaluation were performed under the same conditions. In total, 20 evaluations were carried out: 10 for each assessment. The goal was to execute the four movements of the rehabilitation exercises mentioned above. For each movement, the patient was seated and asked to make three sets of two repetitions each. The MPU-6050 was used in both tests to accurately measure the patient's range of motion.

4.2. Therapist Evaluation

In this test, the therapist guided the patient during the session; the input given by the therapist was verbal, providing a reference with the hand pointing near the specified position, and no additional references were used by the therapist. The patient's movement were detected and processed by the Arduino throughout the session. Neither the therapist nor the patient could see what the Arduino detected at this point. The goal was to use the device to provide exercise feedback to compare with specified foot movements.

The patient was asked to sit to start each session and place his foot on the ground at an angle of $90°$, this first movement was identified as the initial position, which was used to calibrate the device. Also, for each repetition, the patient maintained the foot in the indicated position for five seconds; at this time, the values provided by the accelerometer and gyroscope were taken, processed to generate the angular position, and recorded in Excel. These results were then compared with the results obtained in the evaluation with the application.

4.3. Evaluation with Application

In this test, the application that was developed in Unity provided biofeedback. The application showed the movement, time, series, repetitions, and goals to achieve in that session so that the user could see their performance. The values that were obtained were averaged by movement and stored in the application database so that at the end of the session, the patients could also observe their results and compare them with the expected normal movement range of an ankle.

The information provided by the application at the end of the session, which represents an average of degrees per movement, was recorded and compared with the original information recorded from the therapist evaluation. In this way, we can compare between the results of the feedback provided by the application and the feedback provided by the therapist, which is represented by the difference of degrees between the therapist and the application.

4.4. Results

The values in the degrees recorded from the therapist session and the ones recorded by the application were subtracted. If the difference in degrees was a positive number, it meant that the user had moved her ankle more than the therapist suggested. If the difference was negative, then the user should have moved her ankle further to match the therapist suggestion. It is worth mentioning that the patient performance is not only dependent on how accurately they match the angle given by the therapist. The patients should try to reach the specified angle without hurting the foot, and it is acceptable if the patient moves further than suggested.

Figure 8 shows the averaged difference; root mean square errors of the measured angles were found to be about 1°. Tilting the entire lower leg would affect the ankle joint angular position; the precision of angle measurement depends on how accurately the subject performs the motion. These were then run through the therapist opinions, and were found to be well within the appropriate values. This means that the difference of degrees was not expected to affect the rehabilitation exercises, and movements were done properly using just the feedback provided by the application.

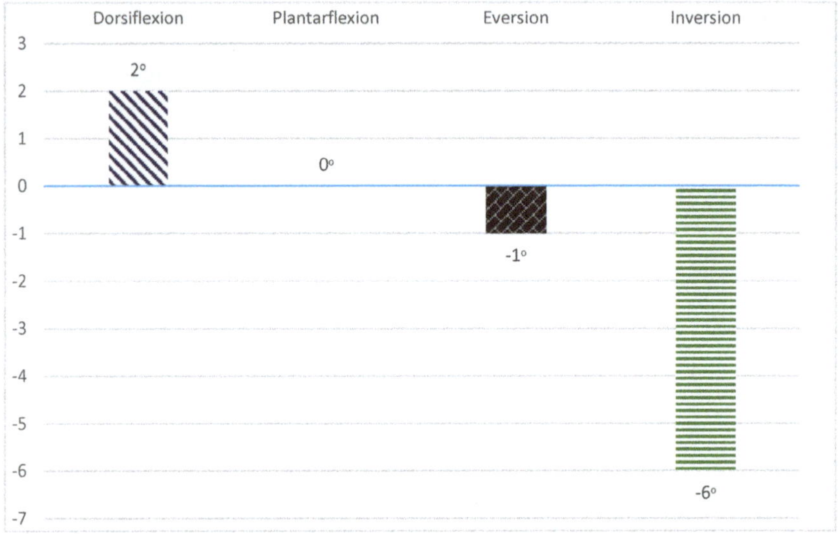

Figure 8. Differences in degrees between software application and therapist.

5. Conclusions

A foot-mounted physical interface for ankle rehabilitation was presented; the proposed solution contains an IMU-based physical interface for measuring the foot attitude, and a graphical user interface that acts as a visual guide for the patient. The physical interface is connected to the computer by a Bluetooth link, and provides feedback to the patient while performing dorsiflexion, plantarflexion, eversion, and inversion exercises. This system allows for in-home rehabilitation at an affordable price while engaging the patient through active therapy. According to the results, more consistent rehabilitation could be achieved by reducing the foot angular position variation during the therapy procedure, measurement differences from 2° to 6° were observed in comparison to a therapist. This is because the therapist process is not as precise as the device. Demonstrative videos are available at the Supplementary Materials section.

Initially, we focused on dorsiflexion, plantarflexion, inversion, and eversion movements using an independent accelerometer (non-IMU). However, to improve the performance, we switched to an IMU MPU-6050, which can take advantage of the gyroscope sensor using data fusion. In future works, abduction and adduction movements could be included. It is also possible to extend the use of this technology to neighboring applications areas such as neurological rehabilitation [27].

Supplementary Materials: The following Video S1: "Demo" is available online at https://www.youtube.com/watch?v=IOvArU4I-yc. Video S2: "Physiotheraphy Ankle" is available online at https://www.youtube.com/watch?v=5Wgqzr2HeLA.

Author Contributions: Conceptualization, N.E.-C. and A.G.-E.; Methodology, A.G.-E.; Software, N.E.-C.; Validation, N.E.-C. and A.G.-E.; Formal Analysis, N.E.-C.; Investigation, N.E.-C.; Resources, N.E.-C. and A.G.-E.; Data Curation, N.E.-C.; Writing—Original Draft Preparation, N.E.-C.; Writing—Review & Editing, B.V.-A. and A.G.-E.; Visualization, N.E.-C.; Supervision, A.G.-E.; Project Administration, B.V.-A. and A.G.-E.

Funding: The authors would like to acknowledge the financial support of Tecnologico de Monterrey, in the production of this work.

Acknowledgments: The authors wish to thank Mauricio Saúl Juárez Rodríguez, and Alejandra Gudiño Rodríguez for their theoretical assistance and valuable suggestions.

Conflicts of Interest: The authors declare no conflict of interest.

References

1. Alcocer, W.; Vela, L.; Blanco, A.; Gonzalez, J.; Oliver, M. Major Trends in the Development of Ankle Rehabilitation Devices. *Dyna* **2012**, *79*, 45–55.

2. English, B. Phases of Rehabilitation. *Foot Ankle Clin.* **2013**, *18*, 357–367. [CrossRef] [PubMed]

3. Meng, W.; Xie, S.Q.; Liu, Q.; Lu, C.Z.; Ai, Q. Robust Iterative Feedback Tuning Control of a Compliant Rehabilitation Robot for Repetitive Ankle Training. *IEEE/ASME Trans. Mechatron.* **2017**, *22*, 173–184. [CrossRef]

4. Rosado, W.M.; Valdés, L.G.; Ortega, A.B.; Ascencio, J.R.; Beltrán, C.D. Passive Rehabilitation Exercises with an Ankle Rehabilitation Prototype Based in a Robot Parallel Structure. *IEEE Lat. Am. Trans.* **2017**, *15*, 48–56. [CrossRef]

5. Sandoval-Palomares, J.J.; Yáñez-Mendiola, J.; Gómez-Espinosa, A.; López-Vela, J.M. Portable System for Monitoring the Microclimate in the Footwear-Foot Interface. *Sensors* **2016**, *16*, 1059. [CrossRef] [PubMed]

6. Nguyen, L.V.; La, H.M. Real-Time Human Foot Motion Localization Algorithm with Dynamic Speed. *IEEE Trans. Hum.-Mach. Syst.* **2016**, *46*, 822–833. [CrossRef]

7. Wang, Z.; Zhao, H.; Qiu, S.; Gao, Q. Stance-Phase Detection for ZUPT-Aided Foot-Mounted Pedestrian Navigation System. *IEEE/ASME Trans. Mechatron.* **2015**, *20*, 3170–3181. [CrossRef]

8. Prieto, J.; Mazuelas, S.; Win, M.Z. Context-Aided Inertial Navigation via Belief Condensation. *IEEE Trans. Signal Process.* **2016**, *64*, 3250–3261. [CrossRef]

9. Ren, M.; Pan, K.; Liu, Y.; Guo, H.; Zhang, X.; Wang, P. A Novel Pedestrian Navigation Algorithm for a Foot-Mounted Inertial-Sensor-Based System. *Sensors* **2016**, *16*, 139. [CrossRef] [PubMed]

10. Bao, S.-D.; Meng, X.-L.; Xiao, W.; Zhang, Z.-Q. Fusion of Inertial/Magnetic Sensor Measurements and Map Information for Pedestrian Tracking. *Sensors* **2017**, *17*, 340. [CrossRef] [PubMed]

11. Shi, W.; Wang, Y.; Wu, Y. Dual MIMU Pedestrian Navigation by Inequality Constraint Kalman Filtering. *Sensors* **2017**, *17*, 427. [CrossRef] [PubMed]

12. Pham, D.D.; Suh, Y.S. Pedestrian Navigation Using Foot-Mounted Inertial Sensor and LIDAR. *Sensors* **2016**, *16*, 120. [CrossRef] [PubMed]

13. Ilyas, M.; Cho, K.; Baeg, S.-H.; Park, S. Drift Reduction in Pedestrian Navigation System by Exploiting Motion Constraints and Magnetic Field. *Sensors* **2016**, *16*, 1455. [CrossRef] [PubMed]

14. Fusca, M.; Negrini, F.; Perego, P.; Magoni, L.; Molteni, F.; Andreoni, G. Validation of a Wearable IMU System for Gait Analysis: Protocol and Application to a New System. *Appl. Sci.* **2018**, *8*, 1167. [CrossRef]

15. Suh, Y.S. Inertial Sensor-Based Smoother for Gait Analysis. *Sensors* **2014**, *14*, 24338–24357. [CrossRef] [PubMed]

16. Tunca, C.; Pehlivan, N.; Ak, N.; Arnrich, B.; Salur, G.; Ersoy, C. Inertial Sensor-Based Robust Gait Analysis in Non-Hospital Settings for Neurological Disorders. *Sensors* **2017**, *17*, 825. [CrossRef] [PubMed]

17. Taborri, J.; Palermo, E.; Rossi, S.; Cappa, P. Gait Partitioning Methods: A Systematic Review. *Sensors* **2016**, *16*, 66. [CrossRef] [PubMed]

18. Chirakanphaisarn, N. Measurement and Analysis System of the Knee Joint Motion in Gait Evaluation for Rehabilitation Medicine. In Proceedings of the Fourth International Conference on Digital Information and Communication Technology and It's Applications (DICTAP), Bangkok, Thailand, 6–8 May 2014. [CrossRef]

19. Moufawad el Achkar, C.; Lenoble-Hoskovec, C.; Paraschiv-Ionescu, A.; Major, K.; Büla, C.; Aminian, K. Physical Behavior in Older Persons during Daily Life: Insights from Instrumented Shoes. *Sensors* **2016**, *16*, 1225. [CrossRef] [PubMed]

20. Ladha, C.; O'Sullivan, J.; Belshaw, Z.; Asher, L. GaitKeeper: A System for Measuring Canine Gait. *Sensors* **2017**, *17*, 309. [CrossRef] [PubMed]

21. Duong, P.D.; Suh, Y.S. Foot Pose Estimation Using an Inertial Sensor Unit and Two Distance Sensors. *Sensors* **2015**, *15*, 15888–15902. [CrossRef] [PubMed]

22. Benoussaad, M.; Sijobert, B.; Mombaur, K.; Azevedo Coste, C. Robust Foot Clearance Estimation Based on the Integration of Foot-Mounted IMU Acceleration Data. *Sensors* **2016**, *16*, 12. [CrossRef] [PubMed]

23. Zhou, Q.; Zhang, H.; Lari, Z.; Liu, Z.; El-Sheimy, N. Design and Implementation of Foot-Mounted Inertial Sensor Based Wearable Electronic Device for Game Play Application. *Sensors* **2016**, *16*, 1752. [CrossRef] [PubMed]

24. Mitschke, C.; Heß, T.; Milani, T.L. Which Method Detects Foot Strike in Rearfoot and Forefoot Runners Accurately when Using an Inertial Measurement Unit? *Appl. Sci.* **2017**, *7*, 959. [CrossRef]

25. Kok, M.; Hol, J.D.; Schön, T.B. An optimization-based approach to human body motion capture using inertial sensors. *IFAC Proc. Vol.* **2014**, *47*, 79–85. [CrossRef]

26. Seel, T.; Ruppin, S. Eliminating the Effect of Magnetic Disturbances on the Inclination Estimates of Inertial Sensors. *IFAC-Pap.* **2017**, *50*, 8798–8803. [CrossRef]

27. Seel, T.; Werner, C.; Schauer, T. The adaptive drop foot stimulator—Multivariable learning control of foot pitch and roll motion in paretic gait. *Med Eng. Phys.* **2016**, *38*, 1205–1213. [CrossRef] [PubMed]

28. Giggins, O.M.; Sweeney, K.T.; Caulfield, B. Rehabilitation exercise assessment using inertial sensors: A cross-sectional analytical study. *J. NeuroEng. Rehabil.* **2014**, *11*, 158. [CrossRef]

29. Shepherd, J.B.; Rowlands, D.D.; James, D.A. A Skill Acquisition Based Framework for Aiding Lower Limb Injury Rehabilitation using a Single Inertial Sensor with Concurrent Visual Feedback. *Procedia Eng.* **2016**, *147*, 632–636. [CrossRef]

30. Long, Y.; Du, Z.-J.; Wang, W.-D.; Zhao, G.-Y.; Xu, G.-Q.; He, L.; Mao, X.-W.; Dong, W. PSO-SVM-Based Online Locomotion Mode Identification for Rehabilitation Robotic Exoskeletons. *Sensors* **2016**, *16*, 1408. [CrossRef] [PubMed]

31. Zhang, H.; Hernandez, D.E.; Su, Z.; Su, B. A Low Cost Vision-Based Road-Following System for Mobile Robots. *Appl. Sci.* **2018**, *8*, 1635. [CrossRef]

32. Borraz, R.; Navarro, P.J.; Fernández, C.; Alcover, P.M. Cloud Incubator Car: A Reliable Platform for Autonomous Driving. *Appl. Sci.* **2018**, *8*, 303. [CrossRef]

33. Elangovan, K.; Krishnasamy Tamilselvam, Y.; Mohan, R.E.; Iwase, M.; Takuma, N.; Wood, K.L. Fault Diagnosis of a Reconfigurable Crawling–Rolling Robot Based on Support Vector Machines. *Appl. Sci.* **2017**, *7*, 1025. [CrossRef]

34. Seel, T.; Raisch, J.; Schauer, T. IMU-Based Joint Angle Measurement for Gait Analysis. *Sensors* **2014**, *14*, 6891–6909. [CrossRef] [PubMed]

35. Seel, T.; Graurock, D.; Schauer, T. Realtime Assessment of Foot Orientation by Accelerometers and Gyroscopes. *Curr. Dir. Biomed. Eng.* **2015**, *1*, 446–469. [CrossRef]

36. "InvenSense Inc". MPU-6000 and MPU-6050 Product Specification Rev. 3.4. Available online: https://store.invensense.com/datasheets/invensense/MPU-6050_DataSheet_V3%204.pdf (accessed on 14 July 2018).

37. "I2Cdevlib". MPU-6050 6-Axis Accelerometer/Gyroscope, I2Cdevlib Device Source and Documentation. Available online: http://www.i2cdevlib.com/devices/mpu6050 (accessed on 14 July 2018).

38. Sabatini, A.M. Kalman-Filter-Based Orientation Determination Using Inertial/Magnetic Sensors: Observability Analysis and Performance Evaluation. *Sensors* **2011**, *11*, 9182–9206. [CrossRef] [PubMed]

39. "MPU-6050 Accelerometer + Gyro". Example Sketch (Base Code). Available online: https://playground.arduino.cc/Main/MPU-6050 (accessed on 14 July 2018).

40. "42Bots". Arduino Script for MPU-6050 Auto-Calibration. Available online: https://42bots.com/tutorials/arduino-script-for-mpu-6050-auto-calibration/ (accessed on 14 July 2018).

 *applied sciences*

Article

Chronic Obstructive Pulmonary Disease Warning in the Approximate Ward Environment

Qing Zhang [1,†], Daniyal Haider [2,†], Weigang Wang [1], Syed Aziz Shah [2,3], Xiaodong Yang [2,*] and Qammer H. Abbasi [3]

[1] Xi'an Jiaotong University Health Science Center, Xi'an 710061, China; tg_xbfnetyy@163.com (Q.Z.); tg_xbfnetyy1@163.com (W.W.)
[2] School of Electronic Engineering, Xidian University, Xi'an 710071, China; daniyalhaider86@gmail.com (D.H.); azizshahics@yahoo.com (S.A.S.)
[3] School of Engineering, University of Glasgow, Scotland G12 8QQ, UK; Qammer.Abbasi@glasgow.ac.uk
* Correspondence: xdyang@xidian.edu.cn
† The authors contributed equally to the work.

Received: 23 August 2018; Accepted: 8 October 2018; Published: 15 October 2018

Abstract: This research presents the usage of modern 5G C-Band sensing for health care monitoring. The focus of this research is to monitor the respiratory symptoms for COPD (Chronic Obstructive Pulmonary Disease). The C-Band sensing is used to detect the respiratory conditions, including normal, abnormal breathing and coughing of a COPD patient by utilizing the simple wireless devices, including a desktop system, network interface card, and the specified tool for the extraction of wireless channel information with Omni directional antenna operating at 4.8 GHz frequency. The 5G sensing technique enhances the sensing performance for the health care sector by monitoring the amplitude information for different respiratory activities of a patient using the above-mentioned devices. This method examines the rhythmic breathing patterns obtained from C-Band sensing and digital respiratory sensor and compared the result.

Keywords: C-Band; Chronic Obstructive Pulmonary Disease (COPD); wireless channel information

1. Introduction

The future of the health sector is very much influenced by the modern 5G wireless sensing technology. The wireless sensing technology is making medical practice very convenient for medical staff by providing remote monitoring facilities to the patients at different locations. In this way, the patient does not need to move from location to location for the monitoring or detection of disease; with the wireless sensing equipment, one can manage and record the blood pressure, heart rate, and breathing pattern of the patient. The 5G sensing using C-band [1,2] is the latest technology which is helpful for any kind of sensing need, especially for the health sector. The health sector is focusing on the usage of Information and Communication Technology (ICT) to improve the patient experience and to minimize the health services cost.

Dysfunction of the cardiorespiratory system can be due to Chronic Obstructive Pulmonary Disease (COPD), arteriosclerosis and asthma. The C-band wireless technology allows the health sector to monitor the patients with these chronic diseases, especially COPD. COPD is the most common respiratory condition involving the airways and characterized by airflow limitation [3]. With COPD, the airways become obstructed and the lungs do not empty properly, leading the air to be trapped inside the lungs. So, people with COPD usually have lower Forced Vital Capacity (FVC). The change in the respiratory system is noticeable by measuring the forced expiratory volume in 1 s (FEV1). In COPD, the FEV1 level becomes lower as compared with FVC. The signs and symptoms of COPD include wheezing, caused by the opening of small airways, hypoxemia (low oxygen level in the blood), hypercapnia (high level of carbon dioxide in the blood), Dyspnea, chronic cough and sputum production and chest tightness. The main cause of COPD is smoking. Other causes include the exposure to different lung irritants, including dust, air pollution etc.

The morbidity rate increases in COPD patients at the age of 40 [4,5]. Worldwide, it is the 4th leading cause of death and it is increasing nowadays. In COPD, maintaining the breathing rate is very important for good health. Respiratory activity is an important parameter for measuring the health of any person [6]. COPD is a progressive disease and with the passage of time without any early detection or diagnosis, it leads to death. Early detection of the disease by careful monitoring of the symptoms reduces the severity of the disease. In COPD, the breathing pattern of a patient becomes worse with the passage of time so there must be some procedure or technique to measure the breathing pattern for diagnosis. The most common breathing monitoring system is invasive monitoring, which requires physical contact with the patient's body.

Non-invasive breathing monitoring is becoming more popular in the health sector by overcoming the drawbacks of traditional invasive systems. Radio frequency techniques is the leading method due to the effective monitoring of very sensitive and small breathing patterns. There are many other techniques for this like Doppler radar, which is used to measure the shift in the signal reflected from the human body [7]. Faith Erden et al. [6], detect the breathing by using IR sensors and Accelerometer signals. In [8] Wenda Li et al. measure the breathing pattern using passive radar system, which is also a non-contact breathing monitoring system. In [9], Abdelnaseer et al. show the usage of Ubi-Breathe for the harnesses of the (received signal strength indicator) RSSI on WiFi devices. Another approach which proposed the idea of using a (channel state information) CSI-based breathing monitoring system is [10] by Liu et al. This proposes the technique for respiratory monitoring and utilizes periodogram for spectral analysis due to this method requiring longer duration for completing the task.

In our work we propose a non-invasive technique to measure the breathing patterns and coughing of COPD patients. This system can monitor the patient in a timely manner and can carefully detect breathing activity and coughing. Our system leverages the usage of wireless devices operating at 4.8 GHz frequency, Omni directional antenna connected with RF signal generator and the network interface card (NIC) for the 5G wireless technology. The breathing activity and coughing induces different imprints for wireless channel information (WCI), which helped in the monitoring of breathing patterns of patients. We also used an invasive sensor for monitoring breathing and compared the results with WCI data.

The organization of the paper is as follows. Section 1 introduces COPD, Section 2 presents the fundamentals of breathing, Section 3 explains the implementation of the proposed system with C-Band, Section 4 mentions the monitoring of breathing and its analysis, Section 5 explains the Spearman's rank correlation and Section 6 explains the conclusion.

Table 1 shows the comparison of different approaches to monitor the breathing activities of the human body. Our approach is more feasible for monitoring respiratory activities by utilizing less equipment and C-Band 5G technology.

Table 1. Comparison between different techniques for monitoring breathing.

S/No.	Techniques for Monitoring Breathing	Explanation	Advantages	Limitations
1	UWB Radar System	Ala Alemaryeen et al. [11] proposed ultra-wide band (UWB) radar based respiratory monitoring system.	Shows less error, more robust to noise.	Complex, Costly.

Table 1. *Cont.*

S/No.	Techniques for Monitoring Breathing	Explanation	Advantages	Limitations
2	Non-invasive capacitive micro Sensor	Nicolas Andre et al. [12] proposed the concept of using capacitive micro sensors and negative temperature coefficient thermistor integrated on a silicon chip to monitor breathing with the help of wireless sensor system.	Useful for larger space, monitoring is easy.	More hardware, time consuming.
3	RSS system	Neal Patwari et al. [13], proposed the concept of monitoring breathing with the help of received signal strength in wireless sensors network.	Reliable detection, less RMS error.	RSS behaves less efficiently in the presence of fading and scattering, performance may decrease due to multipath propagation.
4	Video Monitoring	Ching We Wang et al. [14], proposed the idea of real-time automated infrared video monitoring technique for detection of respiratory activities.	High accuracy rate, robust to many body movements.	Less cost effective, excessive hardware, complex architecture.

2. Fundamentals of Breathing

Breathing sound has a complex nature, which drags it to the higher order statistical analysis (HOSA). For this, the spectral, respiratory and phase components are involved [15]. The specific function known as Bi-coherence is used for this analysis.

$$\gamma_3(s_1, s_2) = \frac{|B(s_1, s_2)|^2}{P(s_1)P(s_2)P(s_1 + s_2)} \tag{1}$$

where the $B(s_1, s_2)$ is represented as bi-spectrum for the process $\{X(n)\}$ and is shown as,

$$B_n(s_1, s_2) = X_n(s_1)X_n(s_2)X_n{}^*(s_1 + s_2) \tag{2}$$

where $X_n(s_i)$, $i = 1, 2$ is called the complex coefficient for the process $\{X(n)\}$ at some frequencies s_i and the $X^*(s_i)$ represents the complex conjugate [15]. On the other hand, $P(s_i)$, $i = 1, 2$ is the representation of the power spectrum at the given frequencies s_i [15]

$$P_n(s) = |X_n(s)|^2 \tag{3}$$

The process phase structure is the defining point for the bi-coherence function. There is the quadrature phase coupling in the signal due to some nonlinearities so through the magnitude of bi-coherence function, $|\gamma_3(s_1, s_2)|$ we can measure the quadrature phase coupling.

The nonlinearities result in the quadrature phase coupling and the reason for the occurrence is the presence of s_1 and s_2 with their corresponding phase φ_1 and φ_2 in the main signal with other frequency component $s_3 = s_1 \pm s_2$ and its corresponding phase $\varphi_3 = \varphi_1 \pm \varphi_2$. The bi-coherence index has a boundary between 0 and 1. According to this, if the magnitude of (Bi-coherence function) BF is equal to 1 then there exists phase coupling between the frequency components and if the magnitude of BF is equal to 0, then there will be 0 coupling [16].

The skewness coefficient calculation is the main process involved in the analysis for the breathing sound. It is represented as,

$$c_3 = \frac{K_3}{\sigma^2} \tag{4}$$

σ^2 Represents the variance and K_3 represents the third order cumulate.

2.1. WCI System Dynamics

WCI has a unique response without the environmental dynamics and the subcarrier a at some given time t represents the channel information denoted as $H_a(t)$,

$$H_a(t) = \sum_{l=1}^{M} \zeta_l e^{-j2\pi \frac{L_1}{\lambda_a}} + e_a(t) \tag{5}$$

Here a represents the subcarriers and M represents the total number of multipath components, ζ_l representing the complex gain for the multipath components, L represents the length for the multipath components and λ_a represents [17] the wavelength for a and given by,

$$\lambda_a = \frac{c}{f_c + \frac{a}{N_{DFT}T_s}} \tag{6}$$

Here c represents the speed of light, carrier frequency is denoted by f_c, T_s represents the total sampling interval and the size of Discrete Fourier Transform is given by NDFT. The noise signal is considered as thermal noise and can be represented as $e_a(t)$. The delays and gains of multipath components have time invariant nature.

2.2. WCI System with Breathing

The time invariance property of multipath components also holds for breathing. Assuming only one multipath component got affected by breathing, from [18] we can write that,

$$\zeta_1(t) = \zeta_1 \left(1 + \frac{\Delta L_1}{L_1} \sin\theta \sin\frac{2\pi b}{60} t + \phi \right)^{-\psi} \tag{7}$$

In Equation (7), ζ_1 gives us the gain of the multipath component (MPC) number 1 and L_1 gives the length for that MPC. The breathing causes the MPC 1 to be displaced or change its position and is represented by ΔL_1. The path loss exponent is represented by ψ and the angle between the EM wave and the subject is represented by θ. The breathing rate is also very important and measured in breath per minute and represented by b and the breathing initial phase is represented by ϕ.

Breathing shows significant effects on the MPC 1 and varies its path length $L_1(t)$, which is represented as,

$$L_1(t) = L_1 + \Delta L_1 \sin\theta \sin\left(\frac{2\pi b}{60} t + \phi \right) \tag{8}$$

The channel information which we showed in Equation (5) will be represented as,

$$H_a(t) = \zeta_1 e^{-j2\pi \frac{L_1(t)}{\lambda_a}} + \sum_{l=2}^{M} \zeta_l e^{-j2\pi \frac{L_l}{\lambda_a}} + e_a(t) \tag{9}$$

Further explanation of Equation (9) is shown as,

$$H_a(t) = \zeta_1 e^{-j2\pi \frac{L_1(t)}{\lambda_a}} e^{-j2\pi \frac{\Delta L_1 \sin\theta \sin(\frac{2\pi b}{60} t + \phi)}{\lambda_a}} + \sum_{l=2}^{M} \zeta_l e^{-j2\pi \frac{L_l}{\lambda_a}} + e_a(t) \tag{10}$$

Jacobi-anger expansion plays an important role in Equation (10) and by decomposing the first term of Equation (10) into finite summation [19], we get,

$$e^{-j2\pi \frac{\Delta L_1 \sin\theta \sin(\frac{2\pi b}{60} t + \phi)}{\lambda_a}} = \sum_{n=-\infty}^{\infty} (-1)^n J_n(v_a) e^{jn\frac{2\pi b}{60} t} e^{jn\phi} \tag{11}$$

The value for $v_a = 2\pi \sin\theta \Delta L_1 / \lambda_a$ and the nth order Bessel function is shown as $J_n(z)$ with argument z.

The practice shows that if the value of $|n| \geq 2$ then the function $J_n(v_a)$ decays in faster manner by giving the ordinary v_a values. With this information our Equation (11) shows the approximation with the values of n as ± 1 and the DC component where $n = 0$. So, the new $H_a(t)$ can be represented as,

$$H_a(t) \approx \underbrace{\zeta_1 e^{-j2\pi\frac{L_1}{\lambda_a}} \sum_{n=-1}^{+1} (-1)^n J_n(v_a) e^{jn\frac{2\pi b}{60}t} e^{jn\phi}}_{U_a} + \underbrace{\sum_{l=2}^{M} \zeta_l e^{-j2\pi\frac{L_l}{\lambda_a}} + e_a(t)}_{TI_a} \tag{12}$$

In the above equation, the U_a represents the signal for the monitoring of breathing activity on the subcarrier a. On the other hand, time invariant part is represented by TI_a, which exists as the response of a static environment and is considered as the interference.

3. Design Implementation and C-Band Sensing

In a communication link, the channel properties must be known in advance, which could be analyzed with the help of wireless channel information (WCI). There exist multiple effects like scattering, fading, and multipath propagation [20], which affect the WCI of RF signal and which propagate between transmitter and receiver. The RF signal could be represented as follows,

$$H(a) = \left\lVert H(a) \right\rVert e^{(b\angle H(a))} \tag{13}$$

In Equation (13) $H(a)$ represents the WCI and the phase and amplitude are represented as, $||H(a)||$ and $\angle H(a)$ respectively. This experiment was performed using C-Band operating at 4.8 GHz frequency. In our research we have used an RF signal generator (DSG 3060) connected to an omnidirectional antenna (Rubber duck) at the transmitter side with an output power of 100 mW (20 dBm) and with typical phase noise of <-110 dBc/Hz at 20 KHz. This antenna propagates the RF signal to 360 degrees horizontally with $\frac{1}{4}$ dipole length of 26 mm. At the receiver side, a desktop mounted with network interface card (MCX416A-BCAT), worked as a receiver. The experiment was performed in different indoor environments. We used microwave absorbing material to reduce the multipath propagation. The WCI then extracted [21] using this NIC mounted in a modified desktop PC. The NIC card revealed only 30 subcarriers' information represented as single WCI packet. After successful pinging of an AP, we received 10 WCI packets per second. The received WCI was in the form of matrix of 30×1 represented as channel frequency response (CFR) and shown as,

$$CFR_{(30x1)matrix} = \left[h^1, h^2, \ldots, h^n\right] \tag{14}$$

In the above equation the h^1 represents the CFR of subcarrier number 1 and h^n represents the CFR for the 30th subcarrier as there are only 30 subcarriers.

Figure 1 shows the experimental settings for the proposed methodology. We have adjusted the distance between the devices and patient according to system feasibility. The height of the transmitter, receiver and patient was adjusted around 1.5 m from the ground and 4.5 m from the ceiling. The total height to the ceiling is 6 m from the ground.

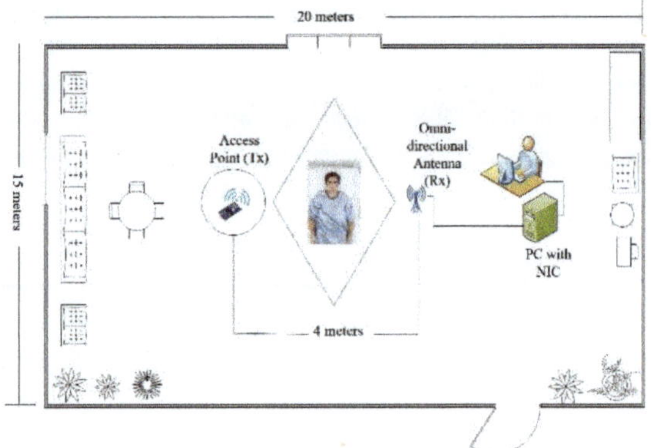

Figure 1. Implemented design overview to monitor breathing.

We obtained the raw WCI for breathing activity of a person imitating a COPD patient. There are 30 subcarriers in total but for the correct breathing information subcarrier #19 and 21 show better results for normal and abnormal breathing and for coughing, subcarrier number 7 gives a reasonable result. We examined all 30 subcarriers [22] and compared their results with our digital respiratory sensor. We performed that experiment multiple times and in all cases the mentioned subcarriers showed better results very near to the digital respiratory sensor's results. In our research we focused on the efficient monitoring of breathing by exploiting the 5G technology. As C-Band is operating on 4.8 GHz frequency, it can be easily integrated with the 5G system. The internet of things (IoTs) operate at a similar frequency band. This way we can easily perform the efficient spectrum utilization regarding 5G. Table 2 shows the numbers of subjects involved in the experiment.

Table 2. Subjects involved in experiment process.

Subjects	Age	Height (cm)	Weight (Kg)
1	29	171	75
2	30	169	78
3	35	161	71
4	31	174	85
5	26	173	87
6	39	176	90
7	40	170	78
8	22	169	73

4. Monitoring Breathing and Its Analysis

This system monitors the breathing activities of a COPD patient [23] which includes normal breathing, abnormal breathing and coughing. We used C-Band sensing technique to extract the WCI for human breathing activities and compared the results with breathing patterns obtained from respiratory sensor HKH-11C. We examined the amplitude information against the time history from WCI using the mentioned subcarriers [22]. Figure 2 shows the raw WCI for coughing.

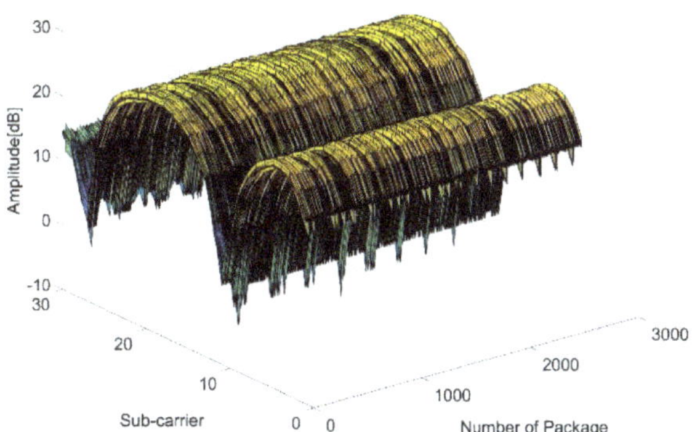

Figure 2. Raw WCI (wireless channel information) for human coughing.

4.1. Monitoring Coughing Pattern

Monitoring coughing in COPD is very important. We used both C- Band sensing and the digital respiratory sensor method. As shown in Figure 2, we obtained the wireless channel information for coughing using C-Band.

Figure 2 shows the amplitude information of all 30 subcarriers. To examine the coughing activity of a COPD patient, we obtained the CFR form C-Band sensing as shown in Figure 3.

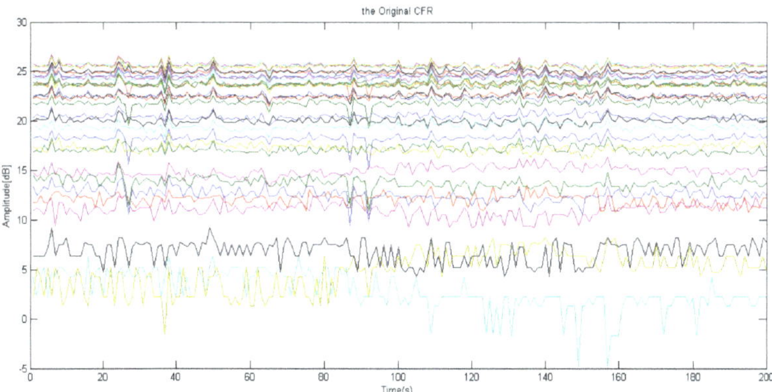

Figure 3. Raw CFR (channel frequency response) for coughing using C-Band.

Figure 3 shows the fluctuation in the amplitude for all the subcarriers against the number of packets. After obtaining this information we performed the filtration process and filter CFRed the raw CFR data by using median filters. The obvious choice for the median and wavelet filter is due to its maximum reduction of noise and smoothing the edges for the available signal.

The above Figure 4 shows the filtered data for the coughing after applying the median filter. The rising and falling edges are much smoother than before with minimum noise effect. This experiment was performed on multiple persons imitating COPD patients, as mentioned in Table 2. We took multiple readings for approximately 1 min each.

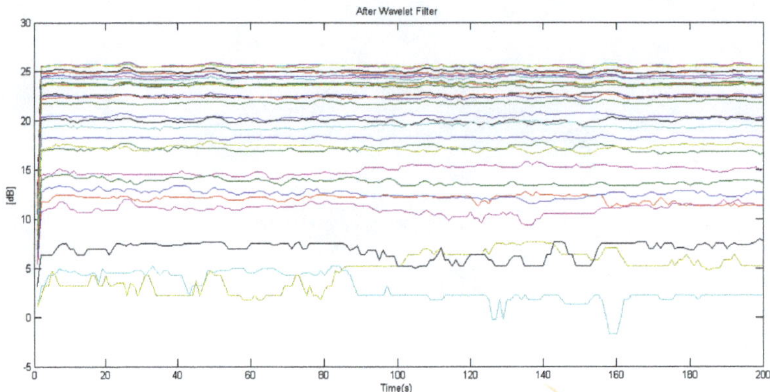

Figure 4. Filtered CFR for coughing.

Now Figure 5 shows the HKH-11C sensor data for coughing; it shows both the raw and filtered data. The person stays in the straight lying position for 60 s wearing the sensor on the chest and imitates the coughing of a COPD patient. Figure 5a shows the raw CFR data for the coughing and Figure 5b shows the filtered data for coughing.

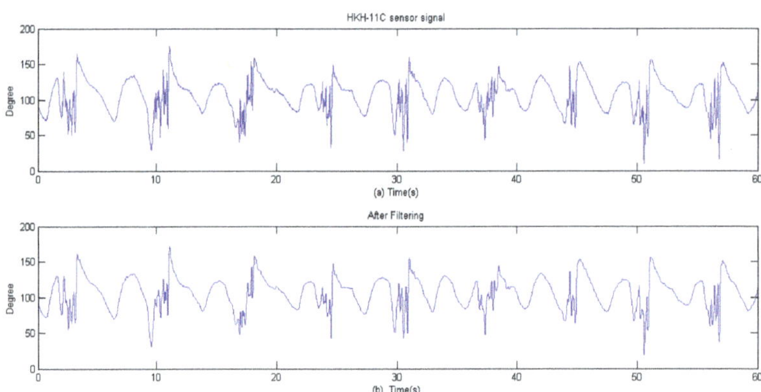

Figure 5. Coughing pattern obtained using HKH-11C sensor. (**a**). Raw coughing pattern. (**b**). Filtered coughing pattern.

The RF signal is always prone to external noise, so frequency selective fading shows greater impact on all the subcarriers. All the 30 subcarriers show different behavior for every power level w.r.t time. As shown in Figure 6, we have selected the sub-carrier # 7 for extracting the finest information for coughing pattern. After obtaining the data from subcarrier # 7, it then passed to the median filter to remove the noise and sharpen the edges.

From Figure 6, we can examine the breathing cycles; the person almost completed nine breathing cycles, as shown in Figure 5. This shows that the result from C-Band and HKH-11C is somehow similar.

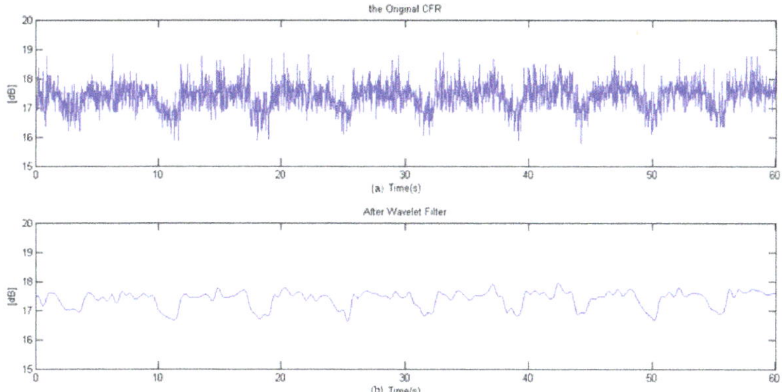

Figure 6. Amplitude information for Subcarrier # 7. (**a**). Raw coughing pattern obtained using C-Band. (**b**). Filtered coughing pattern.

4.2. Monitoring Normal Breathing

In this section we describe the monitoring of normal breathing pattern for a COPD patient. We used 5G C-band technology for extracting the desire WCI. Figure 7 shows the raw WCI for the normal breathing pattern of a COPD patient.

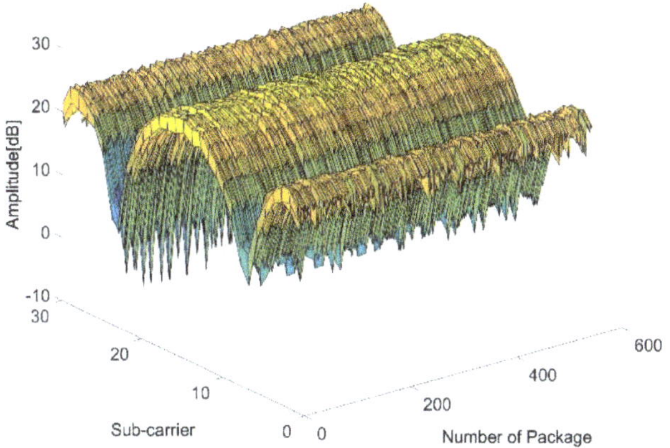

Figure 7. Raw WCI for normal breathing.

To examine the normal breathing activity of a COPD patient, we extract the CFR for all 30 subcarriers using C-Band. The person lying straight imitated the normal breathing for the period of 1 min for multiple times. CFR for all 30 subcarriers for normal breathing is shown in Figure 8.

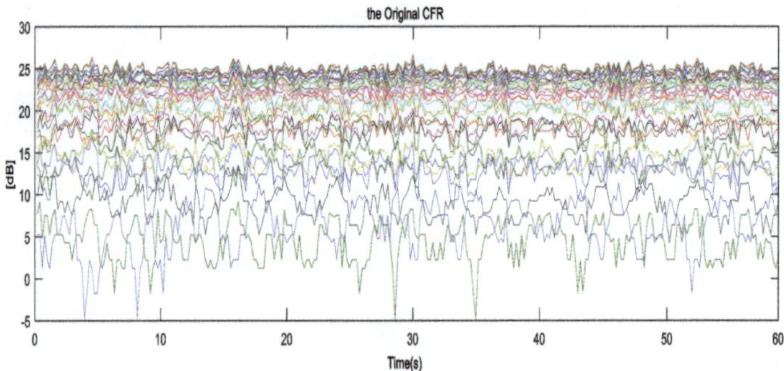

Figure 8. The CFR of all subcarriers for normal breathing.

To obtain the refined CFR we used median filter and wavelet filter to suppress the unwanted noise and smoothen the signal edges. Figure 9 shows the filtered CFR for normal breathing.

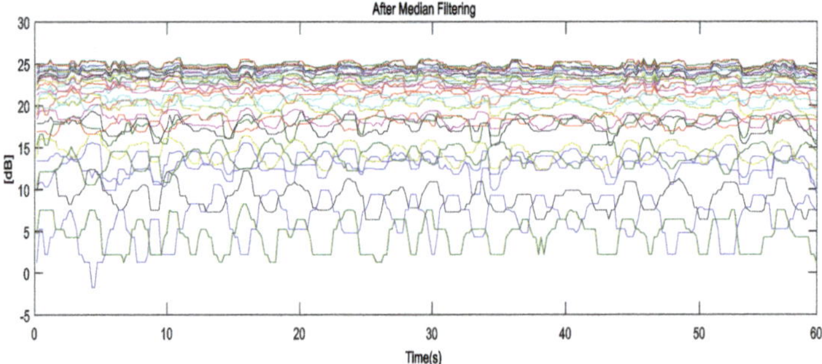

Figure 9. Filtered CFR for normal breathing.

As mentioned before, the experiment was performed on multiple persons lying straight on a bed. We performed this for the period of 60 s each for multiple times to obtain the refined measurement for normal breathing pattern. Figure 10 shows the raw breathing activity using the HKH respiratory sensor for normal breathing. Almost 15 breathing cycles were completed during the period of 60 s.

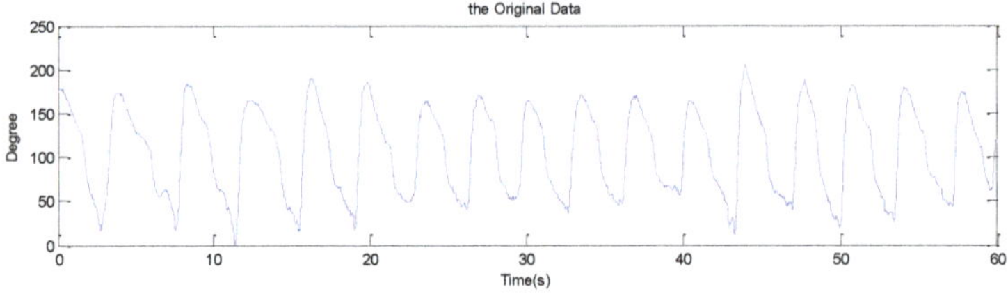

Figure 10. Normal breathing pattern from digital respiratory sensor.

The above Figure 10 shows the raw data; filtered normal breathing pattern is shown in Figure 11. After proper filtration, we obtained the refined normal breathing pattern.

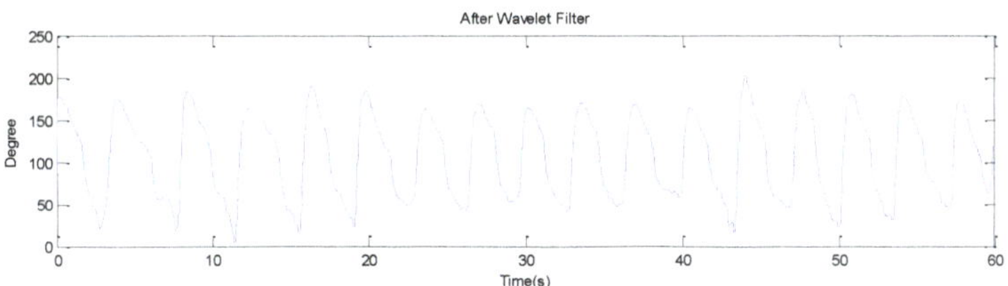

Figure 11. Normal breathing pattern after filtration.

The RF signal shows the clear effects of frequency-selective fading by considering all 30 subcarriers. On different power levels all the available subcarriers show variations w.r.t time. Considering this entire scenario, sub-carrier #19 showed the best results. This subcarrier shows the maximum and adequate information of normal breathing waveform. Figure 12 show the raw WCI for normal breathing using C-Band.

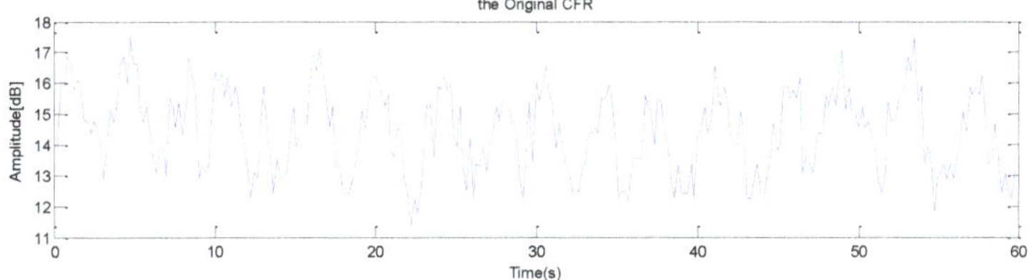

Figure 12. Raw WCI for normal breathing using C-Band.

After the filtration of the above raw WCI, we obtained the refined breathing pattern for normal breathing using C-Band, as shown in Figure 13. Our analysis showed maximum similarity between both the waveform from digital respiratory sensor and C-Band.

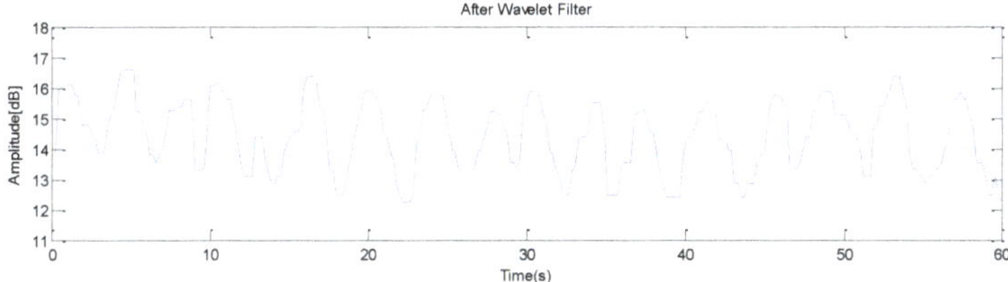

Figure 13. Filtered WCI for normal breathing using C-Band.

From the above Figures 12 and 13, we examine the breathing pattern using 5G C-Band technology. We also examine the fluctuation between the amplitude as a person inhales and exhales the air. The filtration process removes the unwanted fluctuation and smoothens the breathing signal.

4.3. Monitoring Abnormal Breathing

In the last two sections we explained the coughing pattern and normal breathing pattern for COPD patients. We used the C-Band sensing technique and compared the results with the wearable sensor data. We performed the experiment in the same way as we performed the normal breathing experiment; a person imitating a COPD patient was asked to breathe abnormally for a period of 60 s. Through this we collected the raw WCI from all 30 subcarriers with 600 WCI packets. This is shown in Figure 14.

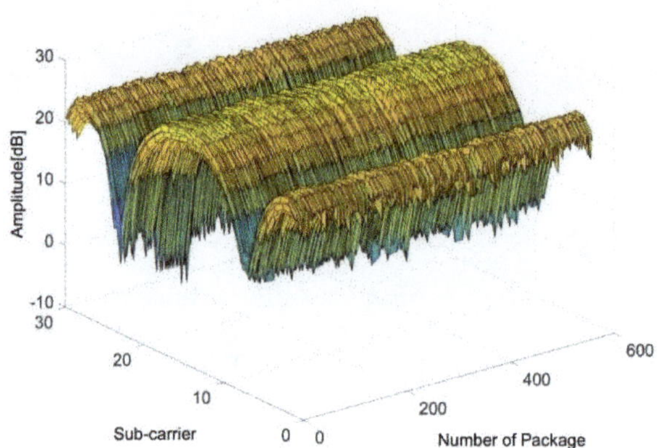

Figure 14. Raw WCI for abnormal breathing.

Now the CFR for all 30 subcarriers are shown in Figure 15. This shows the raw CFR for all 30 subcarriers for the period of 60 s.

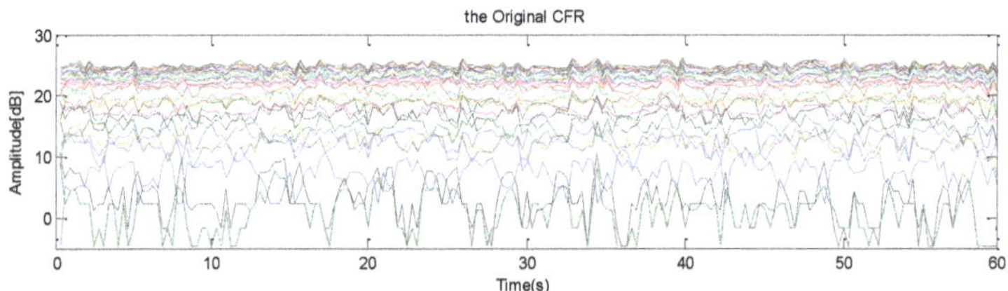

Figure 15. The CFR of all subcarriers for abnormal breathing.

After filtration we obtain the reliable signal for all 30 subcarriers, as shown in Figure 16.

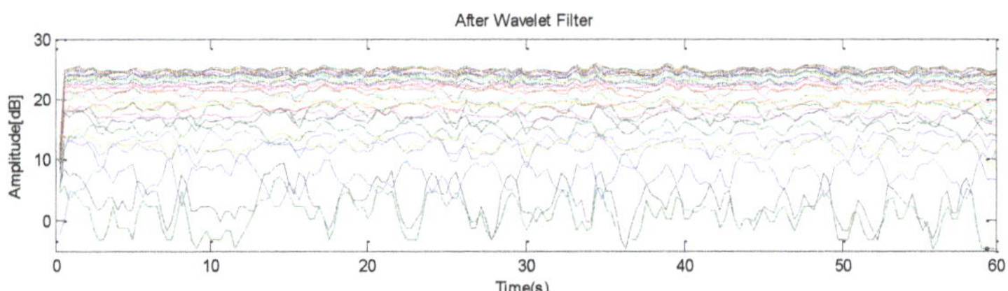

Figure 16. Filtered CFR for abnormal breathing.

We also monitored the abnormal breathing with the HKH-11C breathing sensor, like in the previous section. Multiple people imitated the abnormal breathing, wore the respiratory sensor and inhaled and exhaled the air for a period of 60 s. Figure 17 shows the raw abnormal breathing pattern obtained from the digital respiratory sensor.

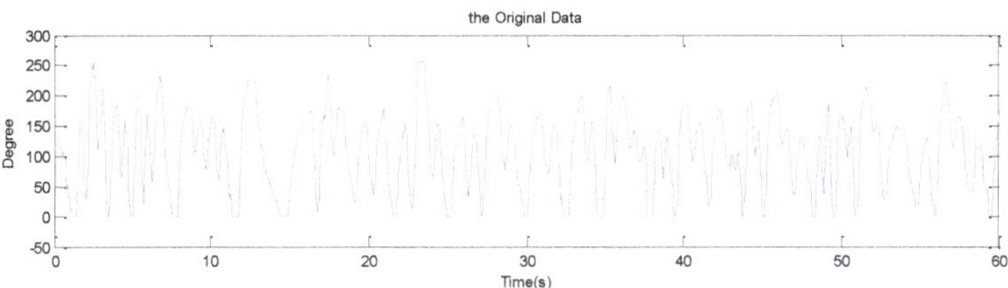

Figure 17. Abnormal breathing pattern from digital respiratory sensor.

The filtered breathing pattern for abnormal breath is shown in Figure 18.

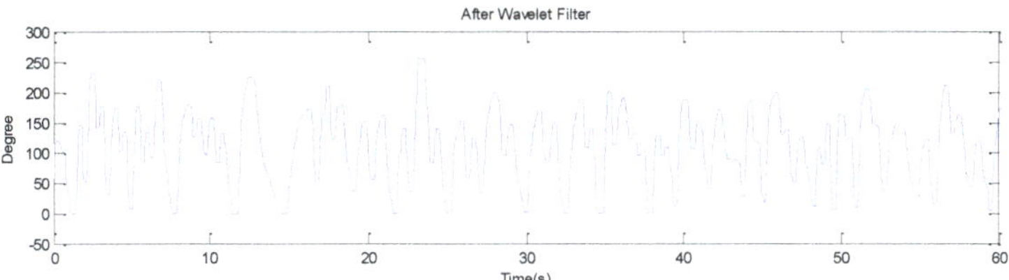

Figure 18. Abnormal breathing pattern after filtration.

The person in the above case imitated the abnormal breathing tracked by the digital respiratory sensor. The breathing sensor behaves as the primary indicator in the above-mentioned case without showing any periodicity. So according to the above Figures 17 and 18 we can see the non-periodicity in the amplitude for both raw and filtered data. This non-periodicity indicates the abnormal breathing pattern.

We also analyzed the data using C-Band utilizing the subcarrier number 21. Figure 19 shows the raw WCI obtained from C-Band. The advantage of using the wireless channel information is that we can choose one or multiple frequency channels for particular applications. The applications can be human intrusion detection, human activity recognition, gait identification and breathing pattern. For each application, specific subcarrier(s) provide the desired information. After examining all 30 frequency channels received for monitoring the breathing pattern of a COPD patient, the subcarriers #7, 19 and 21 gave us the coughing and breathing waveforms. The remaining frequency carriers did not deliver adequate information when the method was implemented in different geometrical locations/experimental settings. We have tested the system in six different locations and at five places the subcarriers #7, 19 and 21 gave us the desired results.

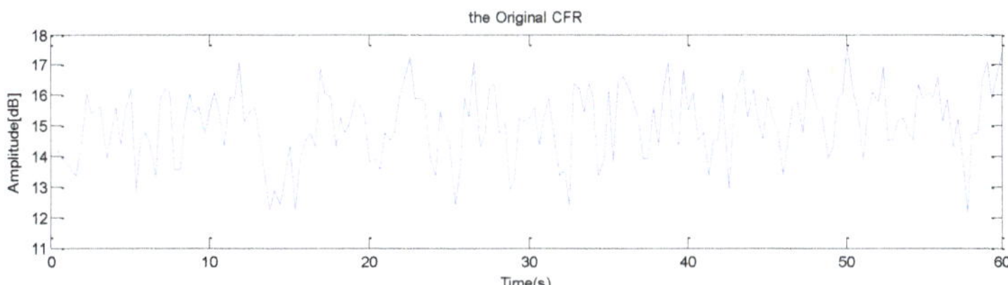

Figure 19. Raw WCI for abnormal breathing using C-Band.

We performed the filtration on the raw WCI obtained from C-Band. After careful examination we concluded that both the waveforms obtained from C-Band and respiratory sensor are almost identical with maximum power level of 18 dB (relative power). In our case we have mentioned the relative power expressed in dB as, $H = Y - X$, where X is the transmitted signal and Y is the received signal. The filtered WCI for abnormal breathing is shown in Figure 20.

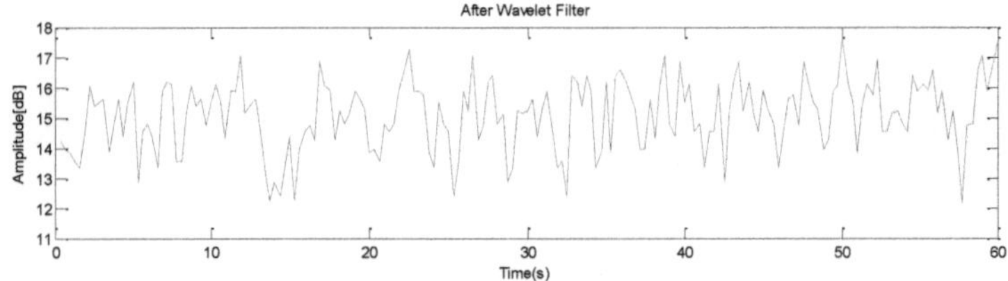

Figure 20. Filtered WCI for normal breathing using C-Band.

5. Spearman's Rank Correlation

Throughout this research our focus revolved around two types of data, one received from C-band sensing and another from the digital respiratory sensor. There exists a correlation between these two types of data and to find that correlation we used the Spearman's rank correlation method. The Spearman's correlation designs itself by utilizing the rank method in which we can obtain the correlation between two signals by analyzing the rank difference. This method reduces the computational time in the presence of a very limited number of observations. The formula for Spearman's correlation is represented as follows,

$$r_k = 1 - \frac{6 \sum D^2}{N^3 - N} \tag{15}$$

r_k = Rank correlation coefficient.

N = Total number of pair observation.

D = Rank difference.

The value for r_k fluctuates between +1 and -1. The direction of rank totally depends on the positive and negative values of r_k. If r_k is positive then rank lies in the same direction and if r_k is negative then rank holds the opposite direction. As mentioned before, we examine two types of signal, one from respiratory sensor and one from C-Band. The r_k value we obtained for normal breathing is 0.89. This value is close to one so accordingly we obtained the positive rank. For the abnormal breathing we obtained the value r_k = 0.86, which is also close to one and again rank shows the positive value. This result shows that the waveforms we captured from both the methods hold some identical features.

6. Conclusions

This paper presented the careful monitoring of breathing activities of a COPD patient using the 5G C-band sensing technique. The obtained results show the feasibility of obtaining wireless channel information from the available setup, where WCI can be used to monitor the coughing, normal and abnormal breathing pattern and then compare it with the data retrieved from the digital respiratory sensor. The presented result was performed on eight different subjects for multiple times lying in a straight position. This research aims to provide an efficient method that can detect the minute chest movement of COPD patients. This method is also very feasible for detecting the heartbeat of a patient to facilitate the doctors and hospital staff.

Author Contributions: Conceptualization, Q.Z., D.H., W.W., X.Y. and Q.H.A.; Methodology, D.H., S.A.S. and X.Y.; Validation, Q.Z. and D.H.; Formal Analysis, D.H.; Investigation, Q.Z. and W.W.; Resources, Q.Z. and W.W.; Data Curation, D.H.; Review & Editing, X.Y.; Project Administration, X.Y.

Funding: This research received no external funding.

Conflicts of Interest: The authors declare no conflict of interest.

References

1. Xu, Y. Emerging Radio Communication Technology and Applications. Article by State Radio Monitoring Center of China Radio Monitoring Department. Available online: https://www.fiercewireless.com/wireless/china-reserves-spectrum-for-5g-says-more-low-band-frequencies-coming-report (accessed on 15 November 2017).
2. Wang, T.; Li, G.; Ding, J.; Miao, Q.; Li, J.; Wang, Y. 5G spectrum: Is China Ready? *IEEE Commun. Mag.* **2015**, *53*, 58–65. [CrossRef]
3. Van Putten, A.F.; Hitchings, D.J.; Quanjer, P.H. Portable electronic peak flowmeter for improved diagnosis of chest diseases in COPD patients. In Proceedings of the IEEE Instrumentation Technology Conference, Irvine, CA, USA, 18–20 May 1993.
4. Garcia-Aymerich, J.; Farrero, E.; Felez, M.A.; Izquierdo, J.; Marrades, R.M.; Anto, J.M. Risk factors of readmission to hospital for a COPD exacerbation: A prospective study. *Thorax* **2003**, *58*, 100–105. [CrossRef] [PubMed]
5. Troosters, T.; Sciurba, F.; Battaglia, S.; Langer, D.; Valluri, S.R.; Martino, L.; Benzo, R.; Andre, D.; Weisman, I.; Decramer, M. Physical inactivity in patients with COPD, a controlled multi-center pilot-study. *Respir. Med.* **2010**, *104*, 1005–1011. [CrossRef] [PubMed]
6. Erden, F.; Cetin, A.E. Breathing Detection Based on the Topological Features of IR Sensor and Accelerometer Signals. In Proceedings of the IEEE 50th Asilomar Conference on Signal and Systems and Computer, Pacific Grove, CA, USA, 6–9 November 2016.
7. Park, B.K.; Yamada, S.; Boric-Lubecke, O.; Lubecke, V. Single channel receiver limitations in Doppler radar measurements of periodic motion. In Proceedings of the 2006 IEEE Radio and Wireless Symposium, San Diego, CA, USA, 17–19 October 2006; pp. 99–102.
8. Li, W.; Tan, B.; Piechocki, R.J. Non-Contact Breathing Detection Using Passive Radar. In Proceedings of the IEEE International Conference on Communication (ICC), Kuala Lumpur, Malaysia, 22–27 May 2016.
9. Abdelnasser, H.; Harras, K.A.; Youssef, M. Ubibreathe: A ubiquitous non-invasive WiFi-based breathing estimator. In Proceedings of the 16th ACM International Symposium on Mobile Ad Hoc Networking and Computing, New York, NY, USA, 22–25 June 2015; pp. 277–286.
10. Liu, J.; Wang, Y.; Chen, Y.; Yang, J.; Chen, X.; Cheng, J. Tracking vital signs during sleep leveraging off-the-shelf WiFi. In Proceedings of the 16th ACM International Symposium on Mobile Ad Hoc Networking and Computing, New York, NY, USA, 22–25 June 2015; pp. 267–276.
11. Alemaryeen, A.; Noghanian, S.; Fazel-Rezai, R. Antenna Effects on Respiratory Rate Measurement Using a UWB Radar System. *IEEE J. Electromagn. RF Microw. Med. Biol.* **2018**, *2*, 87–93. [CrossRef]
12. Andre, N.; Druart, S.; Gerard, P.; Pampin, R.; Moreno-Hagelsieb, L.; Kezai, T.; Francis, L.A.; Flandre, D.; Raskin, J.P. Miniaturized Wireless Sensing System for Real-Time Breath Activity Recording. *IEEE Sens. J.* **2010**, *10*, 178–184. [CrossRef]
13. Patwari, N.; Wilson, J.; Ananthanarayanan, S.; Kasera, S.K.; Westenskow, D.R. Monitoring Breathing via Signal Strength in Wireless Networks. *IEEE Trans. Mob. Comput.* **2014**, *13*, 1774–1786. [CrossRef]
14. Wang, C.W.; Hunter, A.; Gravill, N.; Matusiewicz, S. Unconstrained Video Monitoring of Breathing Behavior and Application to Diagnosis of Sleep Apnea. *IEEE Trans. Biomed. Eng.* **2014**, *61*, 396–404. [CrossRef] [PubMed]
15. Poreva, A.; Karplyuk, Y.; Makarenkova, A.; Makarenkov, A. Detection of COPD's Diagnostic Signs Based on Polyspectral Lung Sounds Analysis of Respiratory Phases. In Proceedings of the 2015 IEEE 35th International Conference on Electronics and Nanotechnology (ELNANO), Kiev, Ukraine, 21–24 April 2015.
16. Ritz, C.P.; Powers, E.J.; Bengtson, R.D. Experimental measurement of three wave coupling and energy cascading. *Phys. Fluids* **1989**, *1*, 153–163. [CrossRef]
17. Chen, C.; Chen, Y.; Han, Y.; Lai, H.Q.; Zhang, F.; Wang, B.; Liu, K.J.R. TR-BREATH: Time-Reversal Breathing Rate Estimation and Detection. *IEEE Trans. Biomed. Eng.* **2018**. [CrossRef] [PubMed]
18. Shah, S.A.; Zhang, Z.; Ren, A.; Zhao, N.; Yang, X.; Zhao, W.; Yang, J.; Zhao, J.; Sun, W.; Hao, Y. Buried Object Sensing Considering Curved Pipeline. *IEEE Antennas Wirel. Propag. Lett.* **2017**, *16*, 2771–2775. [CrossRef]
19. Abramowitz, M. *Handbook of Mathematical Functions, With Formulas, Graphs, and Mathematical Tables*; Dover Publications, Incorporated: Mineola, NY, USA, 1974.

20. Shah, S.A.; Ren, A.; Fan, D.; Zhang, Z.; Zhao, N.; Yang, X.; Luo, M.; Wang, W.; Hu, F.; Rehman, M.U.; et al. Internet of Things for Sensing: A Case Study in the Healthcare System. *Appl. Sci.* **2018**, *8*, 508. [CrossRef]
21. Fan, D.; Ren, A.; Zhao, N.; Yang, X.; Zhang, Z.; Shah, S.A.; Hu, F.; Abbasi, Q.H. Breathing Rhythm Analysis in Body Centric Networks. *IEEE Access* **2018**, *6*, 32507–32513. [CrossRef]
22. Yang, X.; Shah, S.A.; Ren, A.; Zhao, N.; Fan, D.; Hu, F.; Ur-Rehman, M.; von Deneen, K.M.; Tian, J. Wandering Pattern Sensing at S-Band. *IEEE J. Biomed. Health Inform.* **2017**. [CrossRef] [PubMed]
23. Kaltiokallio, O.J.; Yigitler, H.; Jäntti, R.; Patwari, N. Non-Invasive Respiration Rate Monitoring Using a Single COTS TX-RX Pair. In Proceedings of the 13th International Symposium on Information Processing in Sensor Networks, Berlin, Germany, 15–17 April 2014; pp. 59–70.

Article

Movement Noise Cancellation in Second Derivative of Photoplethysmography Signals with Wavelet Transform and Diversity Combining

Dahee Ban, Syed Maaz Shahid and Sungoh Kwon *

School of Electrical Engineering, University of Ulsan, Ulsan 44610, Korea; ekgml3091@gmail.com (D.B.); maaz.shahid26@gmail.com (S.M.S.)
* Correspondence: sungoh@ulsan.ac.kr; Tel.: +82-52-259-1286

Received: 31 July 2018; Accepted: 28 August 2018; Published: 1 September 2018

Abstract: In this paper, we propose an algorithm to remove movement noise from second derivative of photoplethysmography (SDPPG) signals. SDPPG is widely used in healthcare applications because of its easy and comfortable measurement. However, an SDPPG signal is vulnerable to movement, which degrades the signal. Degradation of SDPPG signal shapes can result in incorrect diagnosis. The proposed algorithm detects movement noise in a measurement signal using wavelet transform, and removes movement noise by selecting the best signal from among multiple signals measured at different locations. Experiment results show that the proposed algorithm outperforms the previous filter-based algorithm, and that movement noise with 30% time duration can be reduced by up to 70.89%.

Keywords: SDPPG; movement noise; wavelet transform; diversity combining

---

1. Introduction

With the advancement of electrical and communication technologies, various wearable devices have been developed in a variety of areas including communications, sports, medical diagnosis, and safety. The number of wearable devices will reach up to 485 million in 2018, according to a Business Intelligence report in 2013 [1]. This number is almost 28% of the smart phone market. After the launch of the smart watch, the public has become more interested in wearable devices. In the beginning, most wearable devices were developed for fitness. After smart watch, the application areas expanded from fitness to diverse areas. Healthcare integrating wearable devices with bio-sensing technology are expected to become one of the biggest markets of the Internet of Things [2].

For healthcare, various types of bio-signals are measured such as electroencephalography (EEG), electrocardiography (ECG), electrooculography (EOG), and photoplethysmography (PPG) signals. EEG and ECG are monitoring methods to record electrical activity of the brain and the heart, respectively. EOG is a technique for measuring the corneo-retinal standing potential that exists between the front and the back of the human eye. EEG, ECG, and EOG are measured with a differential amplifier, which registers the difference between two electrodes attached to the skin. These three measurement methods are very uncomfortable for subjects and require knowledge about the human body.

Because of their convenience and portability of measurement, PPG-related signals such as PPG and the second derivative of photoplethysmography (SDPPG) signals are widely used for wearable devices [3]. PPG and SDPPG can monitor heart rates and cardiac cycles by optically detecting blood volume changes in the microvascular bed of tissues. The change in volume caused by the pressure pulse is detected by illuminating the skin with the light from an LED (Light emitting diode), and then measuring the amount of light either transmitted or reflected to a photodiode, as shown in Figure 1 (Near infrared light (600–1000 nm) is suitable for PPG measurement because it permeates the skin and is absorbed into hemoglobin [4]. Therefore, a PPG uses an LED that emits near infrared light to the skin.). For example, more blood absorbs more light so that the receiver receives weaker signal strength. Since PPG and SDPPG only use light signals, they are easier and more comfortable to measure than other kinds of bio-signals. Hence various types of wearable devices with PPG have been developed [5]. However, vulnerability to noise is a major drawback of the PPG-related signals.

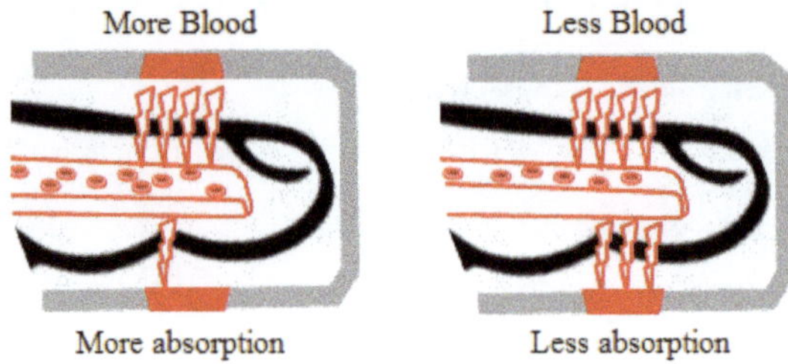

Figure 1. An example of photoplethysmography (PPG) sensing.

In PPG measurement, a signal includes two types of noise: background noise and movement noise [6]. The background noise is caused by electrical noise, such as thermal noise and electromagnetic interference in cables. Movement noise is caused by voluntary or involuntary movements of a subject and affects a wide frequency range. Such noises in measurement distort PPG signal shapes, which can result in wrong diagnoses. Although background noise can be filtered out using a low-pass filter, movement noise cannot be cancelled in this manner [5].

In previous work, there have been efforts to reduce movement noise [6–12]. Patterson et al. [7] studied the best location to measure PPG signals, but movement effect was not considered. Motion artifacts were studied using wavelet transform [8]. Two PPG signals were measured at two different locations with identical PPG sensors. One was used as a reference and the other included movement noise. By comparing the two signals, the properties of movement noises were studied but there was no solution for removal of the movement noise. Kim et al. [6] analyzed the relationship between the accelerometer-measured signal and the PPG signal. Based on this analysis and the least mean squares (LMS), a movement noise cancellation algorithm was proposed. However, the proposed algorithm improves only the peak detection accuracy of PPG signals and needs additional hardware. The adaptive step-size LMS (AS-LMS) adaptive filter [9] was proposed to reduce motion artifacts using a LMS adaptive filter without additional hardware, but the scheme considered only limited cases such as horizontal and vertical movement and therefore cannot be applied for random movement. The periodic moving average filter (PMAF) [10] was proposed to remove movement noise by considering the periodicity of the PPG signals. PMAF segments the PPG signal into periods and resamples each period. The algorithm mitigates a noise at an instant period by averaging multiple signals, but the noise can be propagated to the next signals, which distorts the signal shape. An adaptive spectrum noise cancellation (ASNC) approach for motion artifacts removal in PPG signals is presented by Yang et al. [11]. However, the proposed ASNC utilizes the onboard accelerometer and gyroscope sensors to detect and remove the artifacts adaptively. Peng et al. [12] proposed a method to extract the clean PPG signals from the motion artifacts corrupted PPG signals. The method combined temporally constrained independent component analysis (cICA) and adaptive filters. The cICA extracts the clean PPG signal, and the adaptive filters recover the amplitude information of the PPG signal. But when motion artifacts and PPG signals have dependency, the performance of the proposed method is degraded, i.e., the PPG signal shapes are distorted.

All of the previous studies mainly considered movement noise reduction for measuring heart rates from only PPG signals [6–12] even though SDPPG signals are more useful for diagnosing diseases [13]. The information can be extracted from the signal shapes, so if the shape of SDPPG signals is distorted the applications are limited for diagnosis. Hence, movement noise reduction is important for SDPPG signals.

To remove movement noise generally, and to maintain signal fidelity, we apply signal processing and diversity combining schemes for multiple SDPPG signals. The heart rate is almost periodic with a subject-dependent shape; therefore, a wavelet transform that has multiple bases and changes the time-frequency resolution [14] can detect movement noise effectively. Moreover, the human heart generates a heartbeat whereby PPG signals can be measured at different locations, such as fingers, ears, and toes, as shown in Figure 2. Hence, diversity combining schemes, as with communications, can improve the received PPG signal quality. The quality of each channel is measured by the noise level using the wavelet transform, and the algorithm chooses the best signal. This work can also be applicable to other types of bio-signals that have periodicity and diversity properties for healthcare and disease diagnosis using wearable devices.

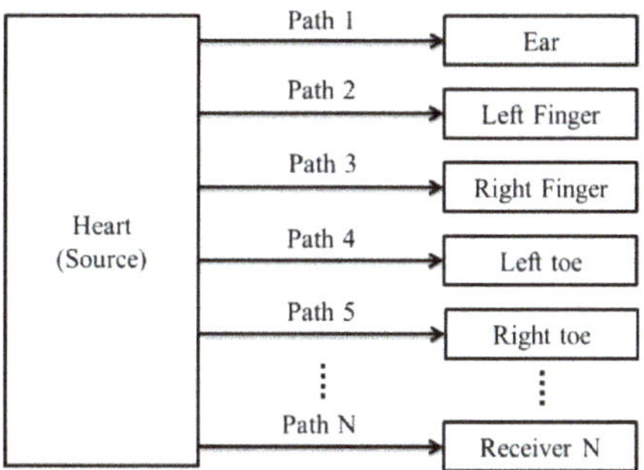

Figure 2. Examples of PPG measurement: ear, fingers, and toes.

Compared to earlier work [15], this paper provides more detail of analysis and combining schemes, improves the proposed algorithm, and verifies the algorithm through more simulations and comparisons. The noise detection using wavelet transform and the noise reduction using a diversity combining scheme are explained in detail. The proposed algorithm is verified by comparing with previous algorithms in various environments.

The rest of this paper is organized as follows. Section 2 describes the properties of SDPPG signals and problems in measurement. Section 3 overviews the wavelet transform and diversity combining for application in analyzing SDPPG signals. The noise reduction algorithm with wavelet transform and selection combining is proposed in Section 4. Section 5 presents experiment results. Finally, this paper is concluded in Section 6.

2. Properties of SDPPG Signals and Problems in Measurement

In this section, we describe the properties of an SDPPG signal and the problems in SDPPG measurement.

2.1. Properties of SDPPG Signals

There are two types of PPG signal that can optically detect blood volume changes in the microvascular bed of tissues, and that are widely used for healthcare: PPG and SDPPG. While PPG directly measures only the stiffness changes in the microvascular bed of the tissue, SDPPG [16] computes the twice-differentiated PPG signal. Since SDPPG emphasizes the inflection points on the original PPG wave, SDPPG provides more information than PPG; so SDPPG is more effective in diagnosing disease.

Figure 3 shows the PPG signal and the corresponding SDPPG signal. An SDPPG signal is characterized by the five sequential waves denoted by a, b, c, d, and e in Figure 3, and the wave information is used to detect vascular diseases. For example, an SDPPG aging index, defined as (b-c-d-e)/a, is associated with atherosclerosis in elderly patients [13]. Therefore, SDPPG is considered instead of PPG in this paper, even though the algorithm is also applicable to PPG.

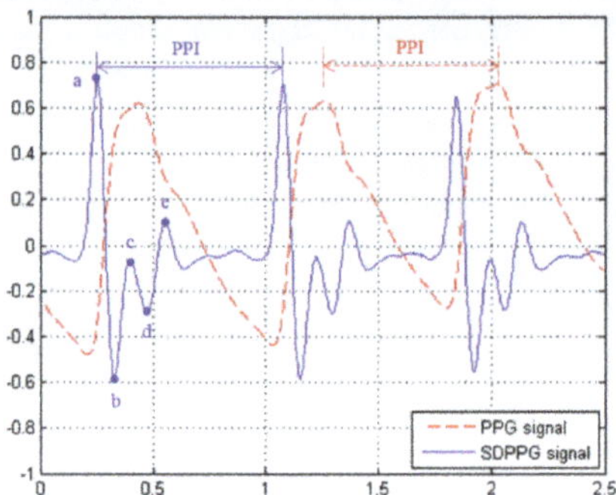

Figure 3. PPG and second derivative of photoplethysmography (SDPPG) signals.

Figure 4 shows the spectrum of PPG and SDPPG signals. The spectrums of PPG and SDPPG are defined between 0 and 10 Hz, and between 0 and 14 Hz, respectively [17], while according to our spectrum analysis the 99% powers of PPG and SDPPG signals are accumulated below 5 Hz and 10 Hz, respectively. Hence, low-pass filtering with cutoff frequency of 20 Hz will not distort the shape of SDPPG signals.

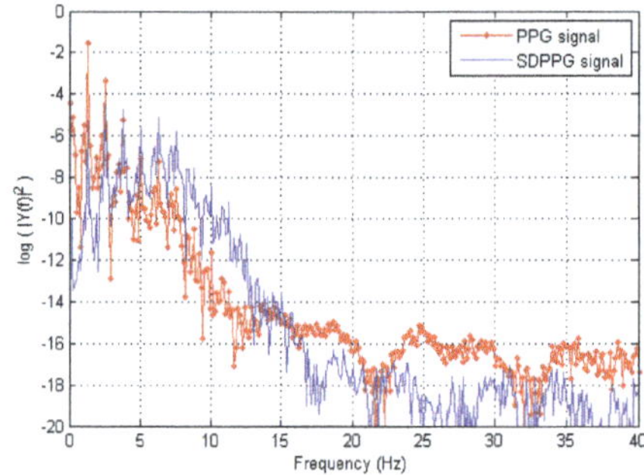

Figure 4. Spectrum of PPG and SDPPG signals.

2.2. Noise in SDPPG Measurement

Using peak to peak intervals (PPIs) of an SDPPG signal, a heart rate can be detected. PPI is defined as the time interval between the maximum values of one signal and the next period signal, as shown in Figure 3. As explained earlier, the SDPPG signal contains not only PPI but also other information that can help diagnose diseases. Hence the shape of the SDPPG signal is crucial. However, noises in measurement such as background noise and movement noise distort SDPPG signal shapes and can result in incorrect diagnoses. Therefore, a noise cancellation algorithm is required to remove the background noise and movement noise.

The background noise maintains the main shape of SDPPG signals, but PPI can be affected. The background noise can be filtered out with a low-pass filter [5]. When there is movement noise in a measurement, the SDPPG signal shapes are significantly distorted, as shown in Figure 5. The movement noise affects SDPPG measurement differently, according to the type of movement, such as walking, vertical motion, or horizontal motion [9]. Hence, it is difficult to analyze and remove movement noise.

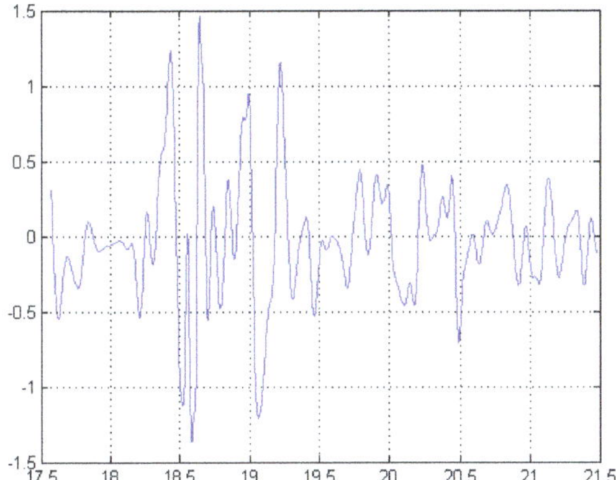

Figure 5. SDPPG signal distorted by movement noise.

In this paper, an algorithm is proposed to detect and remove movement noise in SDPPG signals using wavelet transform and multi-path signals. To that end, we briefly review the wavelet transform and diversity combining in the next section.

3. Overview of Wavelet Transform and Diversity Combining

3.1. Overview of Wavelet Transform

The wavelet transform is one of the most popular methods of time-frequency transformations; it has multiple bases and changes the time-frequency resolution. The heartbeat pulse wave is an almost periodic signal, defined as a signal that is periodic to within any desired level of accuracy. In a normal state, a subject-dependent pulse shape repeats in time, and the shape and period of the pulse can vary slightly. Hence, wavelet transform is suitable for analysis of PPG and SDPPG signals.

The discrete wavelet transform performs decomposition of signals into multi-resolution sub-band representations with two disjoint digital filters: a low-pass filter of impulse response $g[n]$ in (1) and a high-pass filter of impulse response $h[n]$ in (2). The two filters are a quadratic mirror filter, which satisfies the half-band condition [14]. Output signals down-sampled by 2 via the low-pass filter and the high-pass filter are called approximation coefficients ($A[n]$) and detail coefficients ($D[n]$), respectively. Hence, the original signal $S[n]$ consists of the approximation and detail coefficients, as seen in (3).

$$A[n] = (x * g)[n] = \sum_{k=-\infty}^{\infty} x[k]g[2n - k] \tag{1}$$

$$D[n] = (x * h)[n] = \sum_{k=-\infty}^{\infty} x[k]h[2n - k] \tag{2}$$

$$S[n] = A[n] + D[n] \tag{3}$$

The approximation is repeatedly decomposed into two coefficients, approximation (A_N) and detail (D_N) coefficients, where N is the decomposition level of wavelet transform, and the outputs are down-sampled by 2. At the level of the transform, the signal is decomposed into low and high frequencies. Figure 6 shows the example of the 6th level wavelet transform and the corresponding equation is expressed as:

$$S = A_6 + D_1 + D_2 + D_3 + D_4 + D_5 + D_6$$

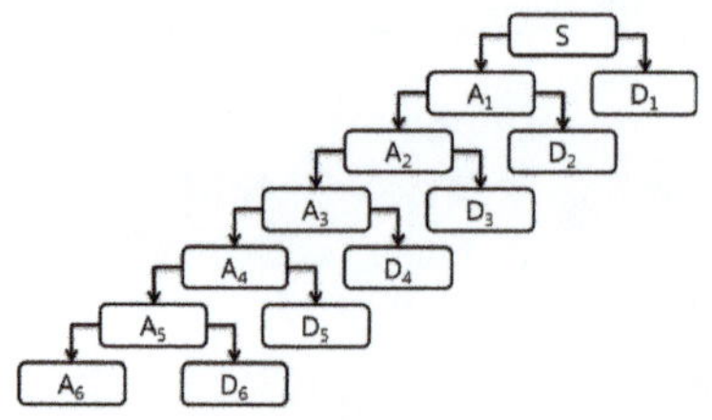

Figure 6. Wavelet packet decomposition.

3.2. Overview of Diversity Combining

A combining technique is popular to combat the effects of distortion, referred to as fading, on wireless systems [18]. When there are a transmitter and multiple receiving antennas, the received signals have different degradation since they experience different fading levels.

The receiver combines diverse signals bearing the same information in order to increase the overall quality of the received signal [18]. There are four types of diversity combining: maximal-ratio combining, equal-gain combining, selection combining, and switch combining. Maximal-ratio combining sums the received signal with weights. Equal-gain combining sums all the received signals synchronously. Selection combining selects the best signal among the multiple received signals. Switch combining switches to another signal when the quality of the selected signal goes below a certain threshold.

In this study, there is only one source (heart) and diverse signals (multiple signals measured at different locations). Hence, a diversity combining scheme is employed to improve the received signal quality. While the signal-to-noise ratio is used as a quality measure in the communication community [18], a signal and noise cannot be separately measured in an SDPPG (or PPG) signal. Hence, for diversity combining, a quality measure for SDPPG signals will be defined and explained later.

4. Proposed Algorithm

In this paper, an algorithm is proposed to detect movement noise and mitigate the impact of mobility on SDPPG measurement by using multipath signals and the wavelet transform. First, SDPPG signals are measured at different locations on a body, such as fingers and ears, and background noise is removed using a low-pass filter. The algorithm determines a reference signal using the measured SDPPG signals. Time offsets between received signals are aligned in order to compare the measured signals. To determine whether the measured signals include movement noise, the wavelet transform is applied to each signal. Based on the composed signals in the frequency domain, the movement noise is detected and removed. The algorithm procedure is summarized in Figure 7 and the details of each step will be explained in the following subsections.

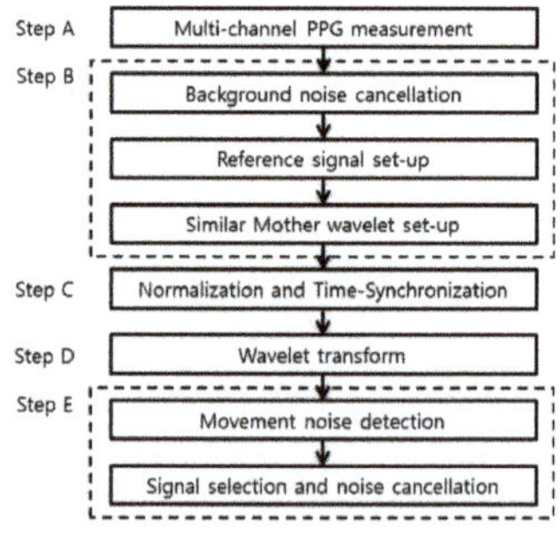

Figure 7. The overall procedure of the proposed algorithm.

4.1. Multi-channel SDPPG Measurement

SDPPG signals can be measured at multiple locations on body details such as fingers, toes, ears, and wrists even if the heart is the only source. Measuring locations that are independently influenced by movement are chosen since the more independent multipath signals bring the better signal combining performance in wireless communications [18]. For example, two fingers on the same hand may be identically influenced by arm motion and have strongly correlated movement noise. Hence, a toe and a finger, or a finger and an ear would be a better selection to measure SDPPG signals.

4.2. Selection of Mother Wavelet

Because of ambient and electric noises, measurement signals inherently include a background noise. The electric noise is generally included in the high-frequency band, and interferes with detection of PPI. Such a noise can be removed by a low-pass filter [6]. Since human heart rates are in the range of 40 to 200 per minute [19] and the 99% power of SDPPG is below 10 Hz described in Section 2, a low-pass filter with a cutoff frequency of 20 Hz can remove such a noise.

After removing background noise from the measured signal, the algorithm determines a reference signal of a single period to select the mother wavelet for wavelet transform. To that end, the SDPPG signal of a few periods is averaged, which is set to a reference signal.

Since an SDPPG signal is dependent on the subject, by cross-correlation between the reference signal and mother wavelets, the algorithm chooses the mother wavelet that is the most similar to the reference signal among mother wavelets such as Harr, Daubechies, or Symlet.

4.3. Normalization and Time-Synchronization

Even if the measured SDPPG signals originate from one source (heart), the received signals may have different amplitudes and traveling times owing to the different paths taken, as shown in Figure 2. The differences in traveling time causes phase differences between received signals, as in Figure 8a. Such amplitude attenuation and time deviation induce errors in detection and decision during signal processing. To adjust the amplitudes, the signals are scaled to have the same amplitude as the strongest signal. To compensate for time differences between signals, cross-correlation between channel signals is performed to obtain the time difference, as shown in Figure 9. The figure has a maximum value at 0.23 s, which means that the two signals had a 0.23-s time difference due to the travelling paths. Based on time difference computation, the algorithm aligns the times of the two signals, as shown in Figure 8b.

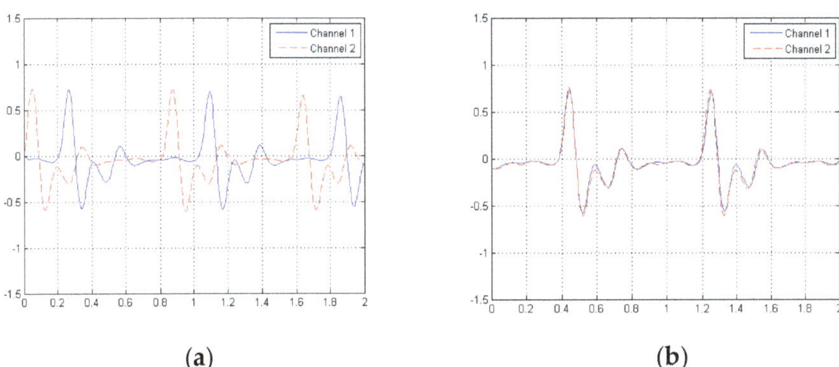

Figure 8. Time-compensation: (**a**) originally measured signals; (**b**) time-aligned two signals.

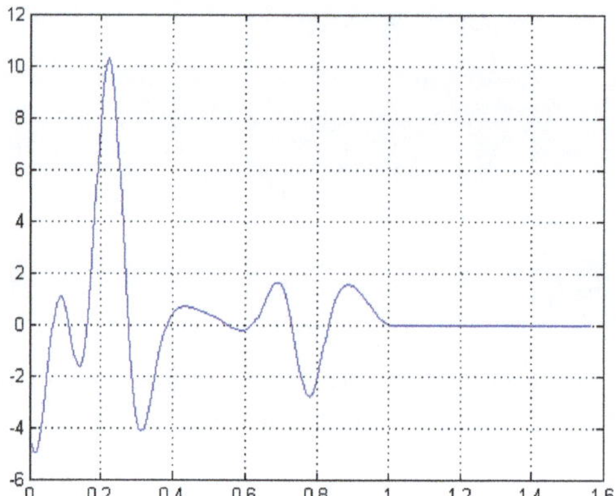

Figure 9. Cross-correlation between two channels.

4.4. Wavelet Transform of SDPPG Signals

With the selected mother wavelet, the wavelet transform is performed at each channel. As described in Section 3, signals are iteratively decomposed up to level N, which depends on the sampling rate, as in Figure 10. For example, if the sampling rate of an SDPPG signal is 256 Hz, then the level N is set to 6 so that the frequency band of the decomposed approximation components becomes between 0 and 4 Hz [14]. Table 1 shows the frequency band according to the level of the wavelet transform at a 256 Hz sampling rate.

Table 1. Frequency Distribution when the Sampling Rate is 256 Hz.

N	A_N (Hz)	D_N (Hz)
1	0–128	128–256
2	0–64	64–128
3	0–32	32–64
4	0–16	16–32
5	0–8	8–16
6	0–4	4–8

A signal period is detected using approximation components, A_N in level N, for which the frequency range is between 0 and 4 Hz, whereas movement noise is detected using detail components D_{N-3} in a level N-3, for which the frequency range is between 32 and 64 Hz. Since measured signals are low-pass filtered out with a cutoff frequency of 20 Hz, as explained in Sections 2 and 4.2, signals higher than the cutoff frequency are mainly related to movement noise at the sampling rate of 256 Hz. In addition, the heart rate is obtained from A_N by computing PPIs.

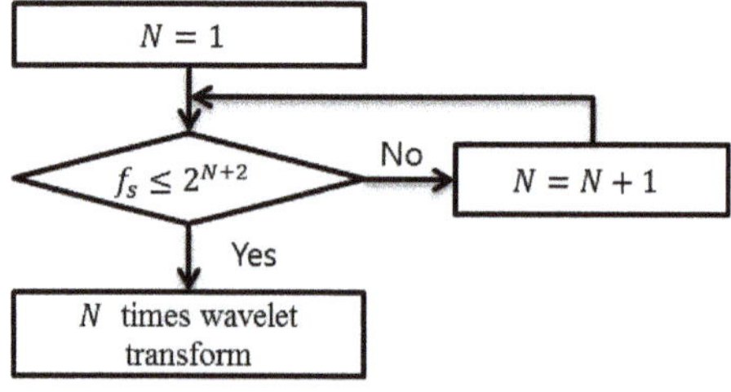

Figure 10. Decomposition level decision procedure.

4.5. Detection and Cancellation of Movement Noise

The algorithm detects noise in SDPPG signals during periods of pulse waves. Instead of continuous time, we consider time slots of which length is the pulse wave period. The length of the time slots can vary according to subjects. Even if signals are measured for a single subject, the duration of the pulse wave can vary according to physical conditions, so a fixed slot time cannot be used for this work. Hence, the algorithm first defines time slots of a pulse wave from the approximation coefficients that contains frequencies between 0 and 4 Hz, and determines whether the movement noise is included in each time period using the detail coefficient that contains a frequency range between 32 and 64 Hz. Each time slot is defined as the time from a peak time to the next peak time of A_N. The procedure for movement noise detection is summarized in Figure 11.

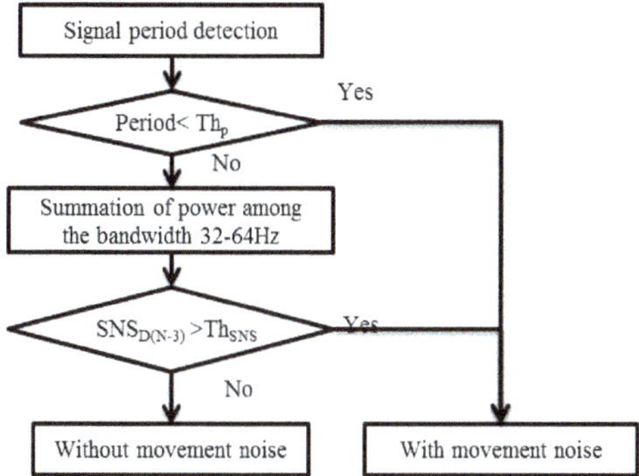

Figure 11. Procedure of movement noise detection algorithm.

After determining a time slot, to decide whether movement noise exists the algorithm checks whether the period of the time slot is less than a threshold (Th_p). In this paper, the threshold is set to 0.3 (In [17], the heart rate is 40–200 beats per minute, which means that the shortest period is 0.3 s.), since the typical heart rate is between 0 and 4 Hz. If the period is less than 0.3 s, i.e., the heart rate is greater than 200 beats per minute, it is decided that the subject is abnormal or the signal is contaminated. If the period of the measured signal is greater than or equal to 0.3 s, the algorithm computes the power of the detail components between 32 and 64 Hz to detect a movement noise.

To determine how much noise is included in a signal for any duration, for a given period the signal-plus-noise-to-signal (SNS) is defined as

$$SNS = \frac{\text{Total power of measured signal and noise}}{\text{Total power of a reference signal}}$$

As a signal and a noise cannot be separately measured, the algorithm is not able to directly use the signal quality used in communications, which is defined as the signal-to-noise ratio. $SNS^i{}_S$ and $SNS^i{}_{D(N-3)}$ denote SNS of the total signal and SNS of the detail component of level $N - 3$ at channel i, respectively.

If the SNS of detail component of level $N - 3$, whose frequency range is between 32 and 64 Hz, is greater than a threshold $Th_{D(N-3)}$, the algorithm decides that the SDPPG signal includes movement noise during the period. Otherwise, the SDPPG signal has no noise. If only one channel has no noise among the channels, the algorithm selects the channel signal at the period without further computation. Otherwise, the algorithm calculates $SNSs$ of the measured signals for all channels and chooses the best channel signal as the selection combining scheme [18]. Since the *SNS* closest to one means that the signal has the highest fidelity to the reference signal, the algorithm chooses the best channel such that

$$k = \mathrm{argmin}_{\{i \text{ in all channels}\}} \left| SNS_S^i - 1 \right|$$

Instead of $SNS^i{}_S$, alternative measures for selecting the channel can be used, such as $SNS^i{}_{D(N-3)}$ of the SDPPG signal instead of its detail components.

5. Experiments

5.1. Experiment Environments

For experiments, SDPPG signals were measured at two different places (a finger and a toe or two index fingers) on five adults using at a sampling rate of 256 Hz. Four 120-s experiments were conducted per person.

For a reference signal, SDPPG signals were recorded at each channel after three seconds to reduce the impact of the initial setting, and five periods of the SDPPG signal were averaged. Subjects sitting in a comfortable chair randomly moved their fingers or their toes during a certain time period. For movement noise detection, the period threshold (Th_p) and the *SNS* threshold (Th_{SNS}) are set to 0.3 and 1.5, respectively.

5.2. Movement Noise Detection and Cancellation Performance

Figure 12a shows a measured SDPPG signal with movement noise and Figure 12b,c show A_6 and D_3 of wavelet transform results, respectively. The algorithm extracts time slots of the pulse waves from the peaks (marked with asterisks) of Figure 12b, and detects movement noise from the power of the detail components of Figure 12d during a slot period. To determine whether movement noise is included in the period, the algorithm computes the *SNS* of the detail component of level 3 and compares it with a threshold. If the *SNS* is greater than 1.5, the SDPPG signal in the period contains movement noise; otherwise, the SDPPG signal during the period has no noise. Detection is performed at all channels.

As shown in Figure 12c, when an SDPPG signal is distorted by movement noise, the detail component surges according to the noise level. Hence, movement noise can be detected.

After detecting movement noise in each period, the algorithm selects the best quality SDPPG signal among channels. If only one channel is unaffected by movement noise, the algorithm selects the signal of that channel. If there is more than one channel that has noise, the channel signal where *SNS* is closest to 1 is chosen.

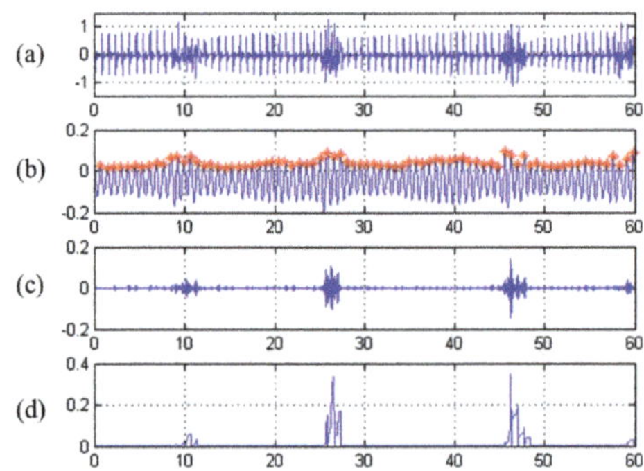

Figure 12. Noise analysis: (**a**) measured signal; (**b**) A_6 components and peaks; (**c**) D_3 components; (**d**) Power of D_3.

Figure 13 shows an example of removing movement noise by the proposed algorithm in various environments. In cases when only one of two channels has movement noise in the time period between 15 and 35 s, as shown in the Figure 13, the noise is removed at the output signals. In cases when all the channels simultaneously have movement noise in the time period between 45 and 60 s, the noise is reduced, but not entirely removed, by selecting the better signal. When two signals are decided to be clean in the time period between 0 and 15 s, the algorithm selects the better signal among them.

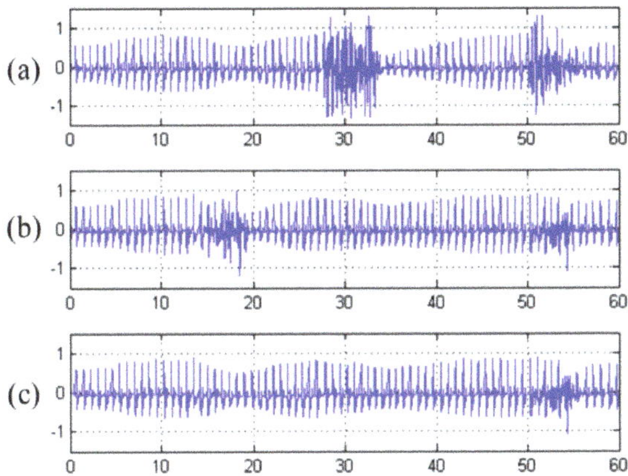

Figure 13. Noise cancellation results: (**a**) an SDPPG signal measured on channel 1; (**b**) an SDPPG signal measured on channel 2; (**c**) an SDPPG signal on output.

We evaluated the performance of our proposed algorithm in terms of accuracy, sensitivity, and specificity (Sensitivity, specificity, and accuracy are used to check the effectiveness of diagnostic test to correctly identify the patients and healthy persons [20]. In our paper, these terms are used to evaluate the performance of our proposed algorithm to correctly identify the noise and noise free time period.). The accuracy is the ratio of the correctly detected noise and noise free period to the total time period. The sensitivity is the ratio of the correctly detected noise period to the total noise period while the ratio of the correctly detected noise free period to total noise free period is define as specificity.

Table 2: Noise detection Performance represents the mean and standard deviation of accuracy, sensitivity and specificity of all cases in channel 2.

Table 2. Noise detection Performance.

	Accuracy	Sensitivity	Specificity
Mean	0.982	1.000	0.977
Standard Deviation	0.010	0.000	0.014

5.3. Performance Comparison

In this subsection, the performance of the proposed algorithm was verified in various movement noise environments, as shown in Table 3. SDPPG signals were measured at pairs of different locations on the subjects: two index fingers (Cases 1, 2, and 3), and one index finger and one toe (Case 4). The subject moved randomly in order to insert movement noise. The noise ratio of channels 1 and 2 in the table implies the ratio of the time period of inserted movement noise to the total experiment time of 120 s. The amount of movement noise on Channel 1 was set to include less than that of Channel 2.

Table 3. Experiment Environments.

	Channel 1		Channel 2	
	Location	Noise (%)	Location	Noise (%)
Case 1	Finger	10	Finger	20
Case 2	Finger	15	Finger	20
Case 3	Finger	30	Finger	30
Case 4	Toe	15	Finger	30

For performance comparison, three previous noise reduction algorithms, PMAF [10], constant step-size LMS (CS-LMS) [9] and independent component analysis LMS (ICA-LMS) are considered. Since these three algorithms are proposed to reduce noise in a single PPG signal, the algorithms are applied to the PPG signal of Channel 1, which contains a smaller amount of noise than Channel 2. After removing movement noise with the previous algorithms, the corresponding SDPPG signals are computed to compare with the proposed algorithm.

Figure 14 shows the experiment results, which are the average of five person results from each case. The noise at output in Figure 14a is the ratio of time period of SDPPG signals with movement noise after noise cancellation to total experiment time. The noise reduction in Figure 14b is the ratio of reduced time period of a noisy SDPPG signal to the time period in which movement noise was inserted on Channel 1.

The experiment results show that the proposed algorithm using wavelet transform and diversity combining can effectively reduce movement noise, and it outperforms the previous algorithms, as shown in Figure 14. While more movement noise in the input signal brings more noise in the results, as shown in Cases 1, 2, and 3 of Figure 14a, the output noise of the proposed algorithm is reduced much more than from the other algorithms, as shown in Figure 14b, even if the second channel signal has more noise than, or the same noise as, Channel 1. For example, in Case 3, the proposed algorithm reduces the amount of noise to 8.73% from 30% (i.e., a 70.89% noise reduction) while PMAF, CS-LMS, and ICA-LMS result in 38.29% (−29.29%), 25.56% (14.79%), and 25.99% (13.38%), respectively. The previous algorithms could not reduce movement noise efficiently while claiming that they improve only PPI variations of a PPG signal [9], [10] and [12], which corresponds to the heart rate variation. Since the PMAF algorithm averages the shapes of multiple periods, the noise can influence the shape of the next periods, which results in deterioration of the SDPPG signals over a greater time duration. Hence, the noise ratios with PMAF increase, and the noise reduction ratios become negative values. When filtering signals, the CS-LMS algorithm partially lessens low-frequency movement noise so that reduction in movement noise is limited. The ICA-LMS algorithm can reduce slow motion artifacts like vertical movement of fingers in heart rates, but it changes and distorts the shape of PPG signals. If the motion artifacts and PPG signals are not independent, the ICA-LMS algorithm performance is degraded, as discussed in [12].

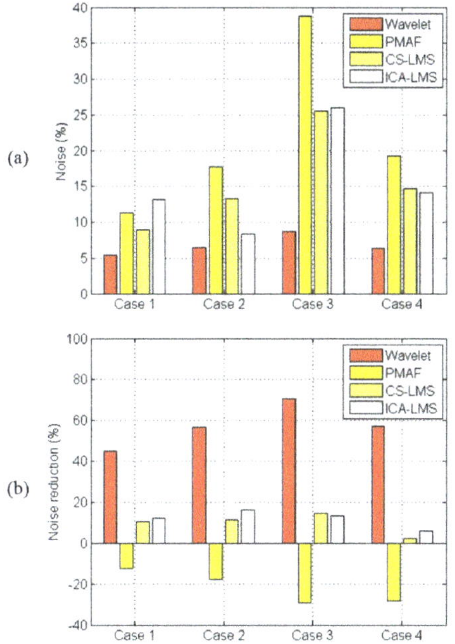

Figure 14. Performance comparison: (**a**) noise ratio at output, and (**b**) noise reduction ratio with respect to the amount of noise on Channel 1.

Our algorithm performance is also robust to measurement locations. In Case 4, which measured signals at a toe and a finger, the performance of the previous algorithms is poorer than in Case 2, which measured two fingers, even if the amount of movement noise on Channel 1 is identical. But our algorithm shows similar performance in Cases 2 and 4. The SDPPG signal measured at a toe has lower power and is more distorted than when measured at a finger due to longer channel path and channel attenuation environment. Hence, the performance of the previous algorithms is degraded, but our algorithm performs robustly at the measurement locations by using wavelet transform and diversity of signals measurement.

6. Conclusions

Among a variety of bio-signals, PPG-related signals such as PPG and SDPPG are widely used for health care applications because of their easy and comfortable measurement. However, PPG and SDPPG signals are vulnerable to movement noise in measurement. Compared to PPG, an SDPPG signal has more information for diagnosing diseases, but is more sensitive to movement noise. Deterioration of SDPPG can result in incorrect diagnosis.

To remove movement noise, an algorithm was developed using wavelet transform and a selection combining technique. Since SDPPG is almost periodic with subject-dependent pulse shape and a band-limited signal, wavelet transform is used for detecting movement noises. SDPPG signals originating from one source (heart) can be measured at different locations so that a diversity combining scheme can be adopted to reduce movement noise. Contrast to previous algorithms, the proposed algorithm improves not only PPI performance but also the shape of SDPPG, which results in better diagnosis. In experiments, movement noise could be reduced by up to 70.89%. The proposed approach is also applicable to PPG and other bio-signals that have periodicity and measurement diversity.

Author Contributions: Conceived the method and wrote the manuscript, D.B.; Designed experiments, D.B. and S.K.; Results analysis, D.B., S.M.S., and S.K.; Writing–review & editing, S.M.S.; Supervision, S.K.

Funding: This work was supported by the 2015 Research Fund of the University of Ulsan.

Conflicts of Interest: The authors declare no conflict of interest.

References

1. Zhang, Y.; Rau, P.L.P. Playing with multiple wearable devices: Exploring the influence of display, motion and gender. *Comput. Hum. Behav.* **2015**, *50*, 148–158. [CrossRef]
2. Hossain, M.S.; Muhammad, G. Cloud-assisted industrial internet of things (iiot)–enabled framework for health monitoring. *Comput. Netw.* **2016**, *101*, 192–202. [CrossRef]
3. Pantelopoulos, A.; Bourbakis, N. A Survey on Wearable Sensor-Based Systems for Health Monitoring and Prognosis. *IEEE Trans. Syst. Man Cybern. C Appl. Rev.* **2010**, *40*, 1–12. [CrossRef]
4. Abe, E.; Fujiwara, K.; Kano, M.; Chigira, H.; Yamakawa, T. Development of Photoplethysmogram sensor-embedded video game controller. In Proceedings of the IEEE ICCE 2016, Las Vegas, NV, USA, 7–11 January 2016; pp. 425–426.
5. Tamura, T.; Maeda, Y.; Sekine, M.; Yoshida, M. Wearable Photoplethysmographic Sensors—Past and Present. *Electronics* **2014**, *3*, 282–302. [CrossRef]
6. Kim, S.; Ryoo, D.; Bae, C. Adaptive noise cancellation using accelerometers for the PPG signal from forehead. In Proceedings of the IEEE Annual International Conference on EMBS 2007, Lyon, France, 22–26 August 2007; pp. 2564–2567.
7. Patterson, J.A.C.; McIlwraith, D.C.; Yang, G.Z. A flexible, low noise reflective PPG sensor platform for ear-worn heart rate monitoring. In Proceedings of the Sixth International Workshop on Wearable and Implantable BSN, Berkeley, CA, USA, 3–5 June 2009; pp. 286–291.
8. Lee, C.; Zhang, Y. Reduction of motion artifacts from photoplethysmographic recordings using a wavelet denoising approach. In Proceedings of the IEEE EMBS Asian-Pacific Conference on Biomedical Engineering 2003, Kyoto, Japan, 20–22 October 2003; pp. 194–195.
9. Ram, M.R.; Madhav, K.V.; Krishana, E.H.; Komalla, N.R.; Reddy, K.A. A novel approach for motion artifact reduction in PPG signals based on AS-LMS adaptive filter. *IEEE Trans. Instrum. Meas.* **2012**, *61*, 1445–1457. [CrossRef]
10. Lee, H.; Lee, J.; Jung, W.; Lee, G. The periodic moving average filter for removing motion artifacts from PPG signals. *Int. J. Control Autom. Syst.* **2007**, *5*, 701–706.
11. Yang, D.; Cheng, Y.; Zhu, J.; Xue, D.; Abt, G.; Ye, H.; Peng, Y. A Novel Adaptive Spectrum Noise Cancellation Approach for Enhancing Heartbeat Rate Monitoring in a Wearable Device. *IEEE Access* **2018**, *6*, 8364–8375. [CrossRef]
12. Peng, F.; Zhang, Z.; Gou, X.; Liu, H.; Wang, W. Motion artifact removal from photoplethysmographic signals by combining temporally constrained independent component analysis and adaptive filter. *Biomed. Eng. Online* **2014**, *13*, 50. [CrossRef] [PubMed]
13. Bortolotto, L.; Blacher, J.; Kondo, T.; Takazawa, K.; Safar, M. Assessment of vascular aging and atherosclerosis in hypertensive subjects: Second derivative of photoplethysmogram versus pulse wave velocity. *Am. J. Hypertens.* **2000**, *13*, 165–171. [CrossRef]
14. Rao, R.; Bopardikar, A. *Wavelet Transforms: Introduction to Theory and Applications*; Pearson Education: Delhi, India, 1998.
15. Ban, D.; Kwon, S. Movement noise cancellation in PPG signals. In Proceedings of the IEEE ICCE 2016, Las Vegas, NV, USA, 7–11 January 2016; pp. 47–48.
16. Takazawa, K.; Tanaka, N.; Fujita, M.; Matsuoka, O.; Saiki, T.; Aikawa, M.; Tamura, S.; Ibukiyama, C. Assessment of vasoactive agents and vascular aging by the second derivative of photoplethysmogram waveform. *Hypertension* **1998**, *32*, 365–370. [CrossRef] [PubMed]
17. Elgendi, M.; Fletcher, R.R.; Norton, I.; Brearley, M.; Abbott, D.; Lovell, N.H.; Schuurmans, D. Frequency analysis of photoplethysmogram and its derivatives. *Comput. Methods Programs Biomed.* **2015**, *122*, 503–512. [CrossRef] [PubMed]
18. Alouini, M.-S.; Marvin, S. An MGF-based performance analysis of generalized selection combining over Rayleigh fading channels. *IEEE Trans. Commun.* **2015**, *48*, 401–415. [CrossRef]

19. Sagie, A.; Larson, M.G.; Goldberg, R.J.; Bengtson, J.R.; Levy, D. An improved method for adjusting the QT interval for heart rate (the Framingham Heart Study). *Am. J. Cardiol.* **1992**, *70*, 797–801. [CrossRef]
20. Zhu, W.; Zeng, N.; Wang, N. Sensitivity, specificity, accuracy associated confidence interval and ROC analysis with practical SAS implementations. In Proceedings of the NESUG Proceedings: Health Care and Life Sciences, Baltimore, Maryland, 2010; Volume 19, p. 67.

Article

An Anonymous Mutual Authenticated Key Agreement Scheme for Wearable Sensors in Wireless Body Area Networks

Chien-Ming Chen [1], Bing Xiang [1], Tsu-Yang Wu [2,3] and King-Hang Wang [4,*

[1]   Harbin Institute of Technology (Shenzhen), Shenzhen 518055, China;
      chienming.taiwan@gmail.com (C.-M.C.); xiangbin.hit@qq.com (B.X.)
[2]   Fujian Provincial Key Laboratory of Big Data Mining and Applications, Fujian University of Technology,
      Fuzhou 350118, China; wutsuyang@gmail.com
[3]   National Demonstration Center for Experimental Electronic Information and Electrical Technology
      Education, Fujian University of Technology, Fuzhou 350118, China
[4]   Department of Computer Science and Engineering, Hong Kong University of Science and Technology,
      Hong Kong, China
*   Correspondence: kevinw@cse.ust.hk; Tel.: +852-2358-8839

Received: 31 May 2018; Accepted: 21 June 2018; Published: 2 July 2018

Abstract: The advancement of Wireless Body Area Networks (WBAN) have led to significant progress in medical and health care systems. However, such networks still suffer from major security and privacy threats, especially for the data collected in medical or health care applications. Lack of security and existence of anonymous communication in WBAN brings about the operation failure of these networks. Recently, Li et al. proposed a lightweight protocol for wearable sensors in wireless body area networks. In their paper, the authors claimed that the protocol may provide anonymous mutual authentication and resist against various types of attacks. This study shows that such a protocol is still vulnerable to three types of attacks, i.e., the offline identity guessing attack, the sensor node impersonation attack and the hub node spoofing attack. We then present a secure scheme that addresses these problems, and retains similar efficiency in wireless sensors nodes and mobile phones.

Keywords: security; anonymity; WBAN; wearable sensors; cryptanalysis

---

1. Introduction

The advancement of electromedical technology has led to new research topics associated with wireless body area networks (WBANs). A wireless body area network (WBAN) is formed by a medication information system and various wearable sensors attached to the patient's body. Integration of WBAN with modern cloud and sensor technologies offers huge improvement in the efficiency and functionality of medical and health care systems. For instance, after the ischemic stroke, patients would require a long-term electrocardiographic monitoring [1]. They suffer from the sleep apnea, and, consequently, require to wear a portable monitor while sleeping [2]. A WBAN-enabled environment allows patients to enjoy the same quality of life without being tangled by the sensor wires. To provide a comprehensive and real-time health assessment to the patient, sensed data may be transmitted to the clouds.

A WBAN architecture is generally constituted of three layers, as shown in Figure 1. This architecture is composed of three types of nodes, first level nodes, second level nodes and a hub node. The first level node, e.g., a smartphone, acts as an intermediate node and forwards the data to the hob node. The second level nodes normally refer to the nodes or wearable devices situated in the body of human, sending the sensing information to a first level node. The hub node a local server or a remote cloud that analyses and manages the sensed data.

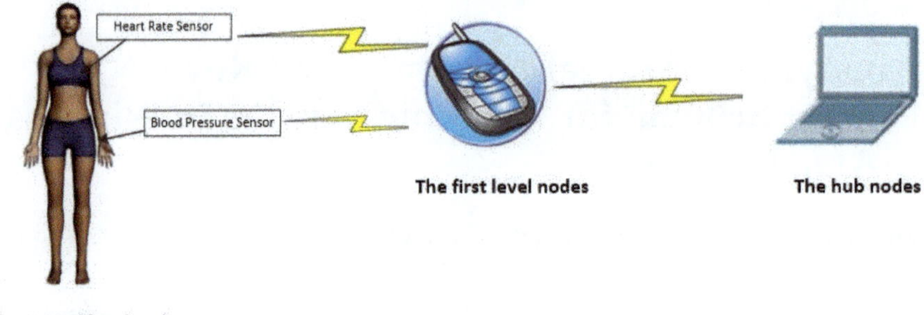

Figure 1. Architecture of a medical WBAN.

Despite the WBANs being endowed with the simplicity and high efficiency, they suffer from low security so that the transmitted data contain the health information of the user which is typically highly sensitive. The need of finding a secure solution for the network is immediate as the security association in the 802.15.6 standard is in doubt [3]. To guarantee a secure WBAN, a secure authentication key agreement protocol should be executed in advance of the communication. We argue that this protocol still requires the user anonymity. Consider a user wearing a portable electrocardiographic monitor to keep track of his cardio health, where the cardio data are appropriately encrypted. The privacy of a known data transfer channel is compromised so that the electrocardiographic monition has been related to a cardio problem through other users.

According to the previously reported works, e.g., [3,4], the authentication key agreement protocol of the WBAN shall provide the data secrecy, user anonymity, session unlinkability, mutual authentication, forward secrecy, resilient to online/offline dictionary attack, resilient to replay attack, and resilient to man-in-the-middle attack. Due to a few reasons, we should not use generic authentication key agreement protocols [5] or lightweight protocols for the general purpose short distance communications [6] in WBANs. Firstly, the specific architecture of the WBAN includes three tiers with multiple first level nodes whose most generic protocols are not optimized in this setting. Some first level nodes may be restricted in terms of power or computation ability so that a heavy computation is not possible. Furthermore, some generic authentication protocols may not offer the user anonymity as their protocol design requirement. However, in a WBAN, the identity of the patient should be concealed while being diagnosed with a WBAN.

WBANs share some similar properties with Hierarchical Wireless Sensor Networks (HWSN). The valuable experience established in the HWSN research area has in turn led to the fast development of WBANs. Wang et al. [7] has summarized some early advancement in the authentication protocol of HWSNs. However, conventional HWSNs assume a large-scale network and are more concerned about the battery power than the security and user's privacy. As of today, there has been no direct applicable of HWSN to WBAN.

Recently, various authentication and key agreement protocols for WBANs have been proposed. In 2009, Keoh et al. [8] has reported a protocol using an on synchronized LED blinking pattern and keychains that provides a visual confirmation of the sensor pairing. Later, Liu et al. [9] presented another protocol using both public key and secret key cryptography in the authentication. In 2014 Liu et al. [10] improved the anonymity over their previous work and presented a protocol focusing on the communication between the first level and second level nodes using the elliptic curve cryptogrpahy and bilinear map. Moreover, the anonymity of the scheme was broken by Zhao in 2014 [11]. Zhao and, subsequently, Wu et al. [12] presented their protocols to overcome some weakness founded in previous works. Those protocols however require the use of public key cryptography (either elliptic curve cryptography or bilinear pairing) in the sensor node yielding a heavy computation and storage bundle [13]. In order to save resources and ensure anonymity, Shen et al. [14] proposed a cloud-aided lightweight authentication protocol. Their protocol ensures that the network manager cannot realize the user's real identity in the authentication phase.

The sensors attached on the human bodies have direct access to the physiological signals of the person. As a result, following the electrocardiogram (ECG) or photoplethysmogram (PPG), the use of these physiological signals may be used to generate keys of the communication [15–17]. Such an approach is quite novel and can be possibly developed in good applications after its robustness and security may be verified in a larger scale or experiments. Unlike secrets, and like passwords or pre-loaded secret keys, the physiological signal may not be necessarily kept away from the attackers.

In 2017, Li and his colleagues proposed a lightweight mutual authentication and key agreement protocol with anonymity for the WBAN [4]. They claimed that their protocol provides anonymity and may be secure against various types of attacks. However, this study demonstrates that Li's protocol is not secure while the first level node is being compromised. In addition, their approach fails to provide the node anonymity so that an attacker is able to track a second level node. To overcome these shortcomings, we provide a simple but effective amendment for the protocol. The repaired protocol is secured against impersonation attacks, replay attacks, and man-in-the-middle attacks. It also provides better anonymity of the WBAN users.

The organization of the paper is as follows. Section 2 reviews the Li's scheme. In Section 3, we show the insecurity of their scheme. Next, an improvement scheme will be presented in Section 4. We then provide some security analysis on the improved scheme, and finally conclude the paper.

2. Review of the Li's Protocol

In this section, we briefly review the Li's protocol [4]. Figure 2 shows the architecture of this protocol, which consists of three level nodes, i.e., a hub node (*HN*), a first level nodes (*FN*) and some second level nodes (*SN*). The second level nodes are some wearable sensors to be attached to the human body. Usually, these *SN* are resource-constrained with limited computational and communicational power. They report sensed data to a first level node (*FN*) via a public channel. A *FN* is an intermediate node between *SN* and *FN*. It may be considered as a smart phone or a smart watch, providing good communication and computation ability and coordinating a set of *SN* attached to the same human body. Next, the *FN* forwards the received sensed data to a hub node (*HN*), which was formed by rich resources and may be installed on a database.

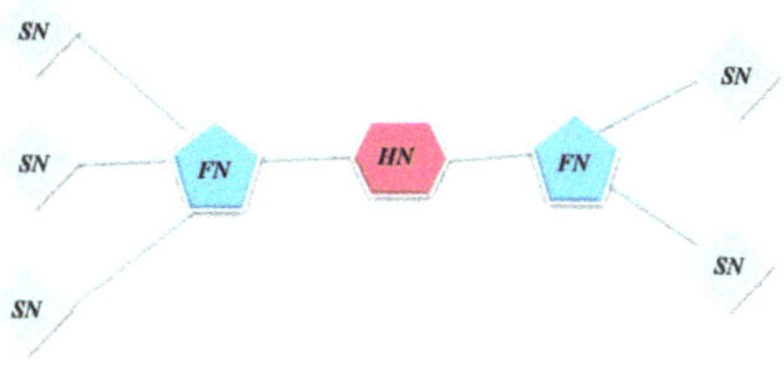

Figure 2. Architecture of Li's protocol [4].

Such a protocol is composed of two phases as follows, the registration phase and the authentication phase. In the registration phase, a system administrator registers and initializes the *HN*, *FN*, and *SN*. In the authentication phase, an *SN* attempts to setup a secure connection in the network while authenticate the identity of the *HN* and being authenticated by the *HN*.

2.1. Registration Phase

In this phase, an *HN* generates a unique secret key, k_{HN}, and securely stores it in its memory. In addition, each second level node is registered individually.

Once a second node *N* is being registered, the following steps are performed:

1. A unique secret identity id_N is generated for the *N* which is also used as the secret key of the *N*.
2. A unique identity id'_N is generated for the *FN*. (It is not explicit in their article that would another id'_N be generated or not when another *SN* is registered. However, if different id'_N is generated for the *SN* that will immediately fail the *SN*'s traceability since the unencrypted id'_N is sent over the air every time the *SN* attempts to connect to the server).
3. A secret parameter k_N is generated for the *N*.
4. The system computes $a_N = id_N \oplus h(k_{HN}, k_N)$ and $b_N = k_{HN} \oplus a_N \oplus k_N$.
5. The *FN* stores the tuple $\langle id'_N, id_N, a_N, b_N \rangle$ in its memory.
6. The *N* stores the tuple $\langle id_N, a_N, b_N \rangle$ in its memory.
7. The *HN* stores the (id'_N) in its memory.

Note that k_N is not required to be stored in the sensor node *SN* or at the hub node *HN*.

2.2. Authentication Phase

In this phase, the *N* establishes a session key with the *HN* through the *FN* as follows. The whole process is given in Figure 3.

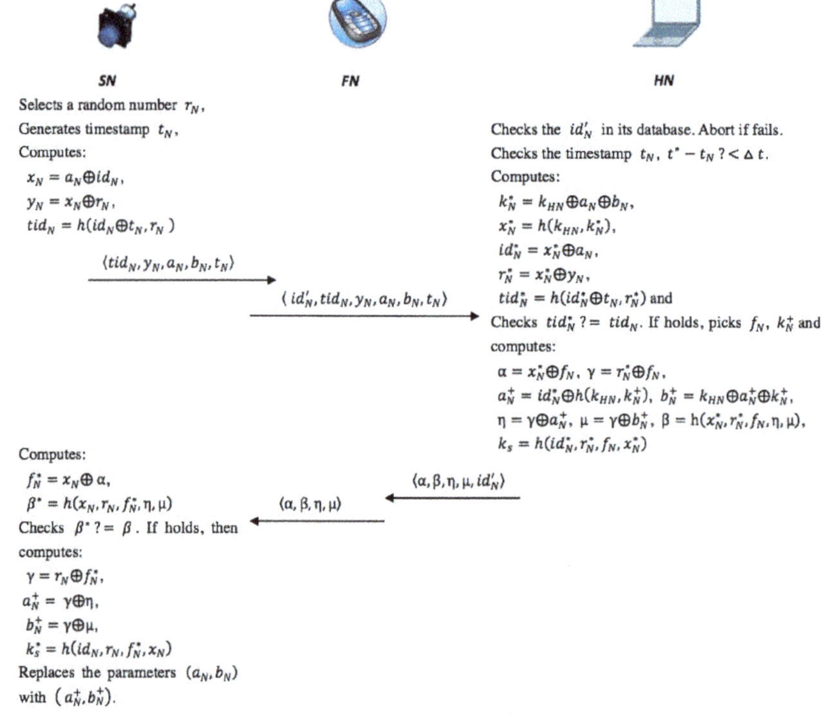

Figure 3. Li's protocol.

1. A second level node N selects a random number r_N and computes

$$x_N = a_N \oplus id_N, \tag{1}$$

$$y_N = x_N \oplus r_N, \tag{2}$$

$$tid_N = h(id_N \oplus t_N, r_N), \tag{3}$$

where t_N is the current timestamp. Next, the N sends $\langle tid_N, y_N, a_N, b_N, t_N \rangle$ to the FN.

2. After receiving the message from the N, the FN places his identity, id'_N, in the message and forwards the message $\langle id'_N, tid_N, y_N, a_N, b_N, t_N \rangle$ to the HN.

3. Once receiving messages from the FN, the HN first checks the id'_N in its database. The process will be terminated if fails. Then, the HN checks the timestamp t_N by judging $t^* - t_N \overset{?}{<} \delta t$, where t^* is the time when the message is received, with δt being the maximum transmission delay. Next, the HN computes the following:

$$k_N^* = k_{HN} \oplus a_N \oplus b_N, \tag{4}$$

$$x_N^* = h(k_{HN}, k_N^*), \tag{5}$$

$$id_N^* = x_N^* \oplus a_N, r_N^* = x_N^* \oplus y_N, \tag{6}$$

$$tid_N^* = h(id_N^* \oplus t_N, r_N^*), \tag{7}$$

4. which checks whether $tid_N^* \overset{?}{=} tid_N$. If the equation holds, the HN ensures that the N is legal. The HN picks temporary secret parameters f_N, k_N^+ and continues to compute the following:

$$\alpha = x_N^* \oplus f_N, \tag{8}$$

$$\gamma = r_N^* \oplus f_N, \tag{9}$$

$$a_N^+ = id_N^* \oplus h(k_{HN}, k_N^+), \tag{10}$$

$$b_N^+ = k_{HN} \oplus a_N^+ \oplus k_N^+, \tag{11}$$

$$\eta = \gamma \oplus a_N^+, \tag{12}$$

$$\mu = \gamma \oplus b_N^+, \tag{13}$$

$$\beta = h(x_N^*, r_N^*, f_N, \eta, \mu). \tag{14}$$

5. Finally, the HN stores the session key $k_s = h(id_N^*, r_N^*, f_N, x_N^*)$ and sends the message $\langle \alpha, \beta, \eta, \mu, id'_N \rangle$ to the FN.

6. Once the FN receives the message from the HN, it drops his identity id'_N and sends the message $\langle \alpha, \beta, \eta, \mu \rangle$ to the N.

7. Now, the N computes $f_N^* = x_N \oplus \alpha, \beta^* = h(x_N, r_N, f_N^*, \eta, \mu)$ and checks $\beta^* \overset{?}{=} \beta$ to determine whether the HN is legal or not. The authentication process will terminate if the equation does not hold. Then, the N computes $\gamma = r_N \oplus f_N^*, a_N^+ = \gamma \oplus \eta$, and $b_N^+ = \gamma \oplus \mu$. Afterwards, the N stores the session key $k_s^* = h(id_N, r_N, f_N^*, x_N)$ and replaces the parameters (a_N, b_N) with the parameters (a_N^+, b_N^+).

3. Cryptanalysis of the Li's Protocol

This section shows that the protocol proposed by Li, and his colleagues, is vulnerable to three types of attacks, i.e., offline identity guessing attacks, sensor node impersonation attacks and hub node spoofing attacks.

3.1. The Adversary Model

We assume the adversary is capable of performing the following, once being attacked. The first three capabilities are adopted from the Li's paper while the last one is a reasonable extension of their model:

- The adversary can control the communication channel. It means that it may eavesdrop, modify and replay any messages transmitted on the communication channel. This intends to capture the protocol requirements, e.g., resilient to replay the attack, resilient man-in-middle attack, mutual authentication, resilient to online/offline dictionary attack.
- The adversary can capture any sensor node by some ways and further extract the secret data store in a captured node. This intends to capture the ability of mutual authentication and forward secrecy.
- The hub node, *HN*, is always trustworthy. However, an adversary may intrude the *HN*'s database and read and manipulate all the data in the database except for the *HN*'s master key, k_{HN}. This intends to capture the resilient of the hub-node-stolen-database attack where the *HN*'s database is stolen.
- An adversary may intrude a first level node *FN* and read all data stored in it. Assuming that both the bottom level *SN* and the top level *HN* can be compromised by the adversary, the *FN* may not remain unintruded for all the time, especially an *FN* may be viewed as a smart phone or a smart watch which may be easily stolen.

3.2. Vulnerable against Intruding FN Attacks

In the protocol design, an *FN* is mainly served as a intermediate relay. However, during the registration phase, the secret information, e.g., id_N, a_N and b_N are all stored in the *FN*. It is not explicit how these values shall be used in the *FN* according to their paper. It is observed that the *FN* does not have the capability to authenticate an *SN* and to be authenticated by the *HN* on behalf of an *SN*, if the *FN* is responsible to coordinate the *SN*. Nevertheless, this turns out to become a point of vulnerability of the protocol. For an adversary which is able to intrude an *FN*, all *SN* s coordinated by this *FN* are compromised.

3.3. Vulnerable to the Tracking Attack

Li claimed that the protocol allows anonymous communication so that an adversary cannot link any communication session to another session of the same *SN*. However, this claim is not true, based on the following facts.

Every *SN* is registered to the system through one single *FN*. The identity of the *FN*, id'_N, is sent over the air in Step 2 of the authentication phase. Since id'_N would not be changed in the protocol, adversary can be easily associated with two sessions with the same *FN* s. For an *FN* coordinating only one *SN*, the adversary is allowed to link two sessions of the same *SN* by inspecting only Step 2. If the *FN* coordinates more *SN* s, the user's privacy/anonymity does not enhance as in some applications suggested in Li's paper. Consider the medication, where the sensors of a patient are likely to be connected to a single *FN*, e.g., his smart phone. Revealing the identity of the *FN* (smart phone) is even worse than revealing only the identity of an *SN* (a sensor).

In certain applications, an *FN* may coordinate extremely large amount of *SN* s, where the identity of the *SN* is the only concern and an adversary is still able to link two sessions with the same *SN* s. Assuming that the adversary $\mathcal{A}$ captures only the messages sent from the *SN* to *FN* and *FN* to *SN* at the time T_1 and a later time T_2, as

$$\text{Capture at } T_1 : \begin{cases} \langle tid_1, y_1, a_1, b_1, t_1 \rangle \\ \langle \alpha_1, \beta_1, \eta_1, \mu_1 \rangle \end{cases} , \quad \text{Capture at } T_2 : \begin{cases} \langle tid_2, y_2, a_2, b_2, t_2 \rangle \\ \langle \alpha_2, \beta_2, \eta_2, \mu_2 \rangle \end{cases} .$$

To investigate if the messages captured at T_2 is a subsequent login of the messages captured at T_1, the $\mathcal{A}$ simply computes $a_2 \oplus b_2$. If these two sessions are related, this value corresponds to $(\gamma_1 \oplus \eta_1) \oplus (\gamma_1 \oplus \mu_1) = \eta_1 \oplus \mu_1$, which is indeed $k_{HN} \oplus k_N$. Except for an extreme low probability of coincident $(2^{-\text{length}(k_{HN})})$, comparing $a_2 \oplus b_2 \overset{?}{=} \eta_1 \oplus \mu_1$ will allow for determining if these two sessions are related.

4. Repairing the Protocol

One of the biggest problems associated with the protocol is that the *FN* does not perform its function in the authentication while it is possessing the secret information of the coordinating *SN*. A simple straightforward approach is to let the *FN* not store any information about the *SN*. Instead, the *FN* only acts as a relay between the *SN* and the *HN*. The protocol will be remaining secure (but not anonymous) even if the *FN* is being compromised. This however does not resolve the vulnerability of the protocl against the tracking attacks. Moreover, this option removes the ability of an *FN* to control other *SNs*, which may not be suitable in some applications.

The security and system requirements may be investigated as follows. The *SNs* assume low computation/communication power; while *FNs* and *HNs* are less constrained, the *SNs* and *HNs* require being mutually authenticated. The *SN* and *FN* should be mutually authenticated where these two authentications may not be necessarily at the same time. Based on these requirements, we propose a simpler repaired protocol exhibiting better security and anonymity.

4.1. Architecture

In our architecture, we maintain the three-level role. However, the communication between an *SN* and an *FN* (*SN-FN*) is different from the communication session between an *SN* and an *HN* (*SN-HN*). A two-party authentication protocol will be described in this section, and the *same protocol* will be used in the case of *SN-FN* and *SN-HN*. In the case of an *SN-HN* communication, the *FN* will be served as a relay to support the communication. The *SN-HN* communication normally takes place when the sensing data is reported to the *HN*. The *SN-FN* communication normally takes place when *FN* manages the *SN* or gathering data from the *SN*. In the case where *FN-HN* communication is required, we assume that general purpose authentication protocols, e.g., [5,18], will be used since both of them have less constraint computation power.

4.2. Description of the Repaired Protocol

As mentioned above, this protocol is a two-party protocol. The reader may assume a duplication of keys for the *SN-FN* and *SN-HN* communications. We call the *UN* an upstream node that represents either an *FN* or an *HN*. Unless it is specified, all variables have the same length as the output of a hash function length(h).

A *SN* should separately register with an *FN* and an *HN*, and two sets of keys are required. Practically, these two registrations may be simultaneously performed via the *FN*, as long as the process is securely accomplished. Assume that the *SN* is registering with either of them, denoted as a *UN*. The *SN* will then be assigned with the followings:

- id_N, a unique secret identity for the *SN*.
- $a_N = id_N \oplus h(k_{UN}, k_N)$, where k_{UN} is the secret key of the *UN*, k_N is a nonce.
- $b_N = a_N \oplus k_{UN} \oplus k_N$.
- $c_N = h(id_N, k_{UN})$.

In this protocol, the *UN* does not require storing any secret information about the *SN*. If the *UN* wishes to keep track of the identity of the *SN*, it may keep a truncated or hashed id_N. The value of the id_N needs to be unique and a bit of id_N may be used to indicate the association with either of *SN-HN* or *SN-FN*, and several bits from the identity of the *UN*.

When the *SN* wishes to initiate a communication with a *UN*, the *SN* will perform the following operations (In case an *FN* wishes to initiate the protocol, the protocol will be preceded by a *Hello* message from the *FN* to the *SN*.). Please also refer to Figure 4.

1. The *SN* generates a random number r_N and a timestamp t_N and computes:

$$x_N = a_N \oplus id_N, \tag{15}$$
$$y_N = x_N \oplus r_N, \tag{16}$$
$$tid_N = h(id_N, t_N, c_N, r_N). \tag{17}$$

Then, it sends $\langle tid_N, y_N, a_N, b_N, t_N \rangle$ to the *UN*.

2. On receiving the request, the *UN* first checks if the timestamp is still valid. Then, it computes:

$$k_N^* = k_{UN} \oplus a_N \oplus b_N, \tag{18}$$
$$x_N^* = h(k_{UN} \oplus k_N^*), id_N^* = x_N^* \oplus a_N, \tag{19}$$
$$r_N^* = x_N^* \oplus y_N, c_N^* = h(id_N^*, k_{UN}). \tag{20}$$

Next, it validates tid_N by $h(id_N^*, t_N, c_N^*, r_N^*)$. The protocol will be aborted if this does not hold.

3. The *UN* continues the protocols by selecting random numbers f_N, k_N^+ and computing the following:

$$a_N^+ = id_N^* \oplus h(k_{UN}, k_N^+), \tag{21}$$
$$b_N^+ = a_N^+ \oplus k_{UN} \oplus k_N^+, \tag{22}$$
$$\eta = h(f_N, c_N^*) \oplus a_N^+, \tag{23}$$
$$\mu = h(c_N^*, f_N) \oplus b_N^+, \tag{24}$$
$$\alpha = c_N^* \oplus f_N, \tag{25}$$
$$\beta = h(id_N^*, r_N^*, f_N, \eta, \mu), \tag{26}$$
$$k_s = h(id_N^*, r_N^*, f_N, x_N^*), \tag{27}$$

where k_s represents the session key. Finally, the *UN* sends $\langle \alpha, \beta, \eta, \mu \rangle$ to the *SN*.

4. The *SN* validates the message by computing $f_N^* = c_N \oplus \alpha$ and checking whether β equals to $h(id_N^*, r_N, f_N^*, \eta, \mu)$. If not, it rejects the protocol.

5. Finally, the *SN* computes the session keys and updates its keys, as

$$a_N^+ = h(f_N^*, c_N) \oplus \eta, \tag{28}$$
$$b_N^+ = h(c_N, f_N^*) \oplus \mu, \tag{29}$$
$$k_s^* = h(id_N, r_N, f_N^*, x_N). \tag{30}$$

The *SN* will compute the same session key k_s as the *UN* in the absence of the adversary or noise. It will then replace (a_N, b_N) with (a_N^+, b_N^+) in its memory.

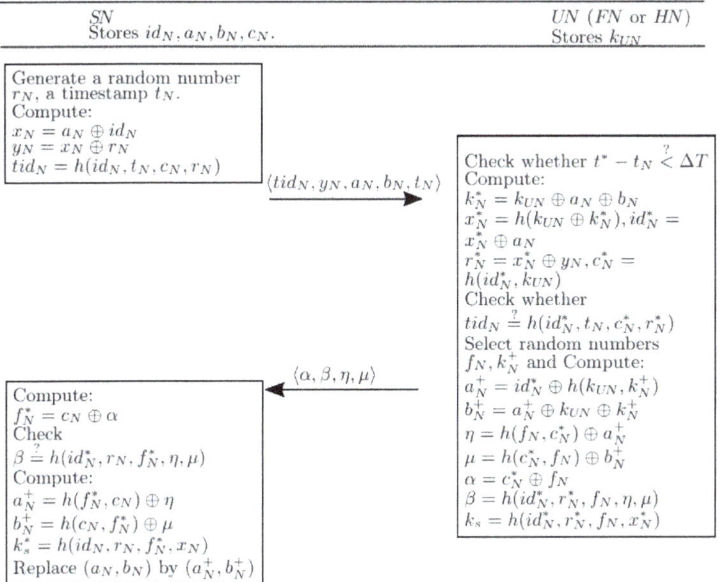

Figure 4. The repaired protocol.

5. Security Analysis of the Repaired Protocol

This section demonstrates that our repaired protocol is secure against the aforementioned attacks.

5.1. Intruding on the FN Attacks

In the repaired protocol, the *FN* no longer stores the key between an *SN* and an *HN*. Therefore, compromising an *FN* would only leak the keys between the *SN*s and the *FN*. The compromised *FN* would not be able to impersonate an *SN* to communicate with the *HN*. It is true that the compromised *FN* will still be able to access the *SN* in an *SN-FN* communication, but no extra access, e.g., data exclusive for the *HN*, will be given to the *FN*. This protocol also assures a secure *SN-FN* communication, and vice versa if all secrets stored in the *HN* are compromised.

5.2. Impersonation, Man-in-the-Middle and Replay Attacks

The protocol provides a sound mutual authentication between an *SN* and an *FN/HN*. The adversary defined in Section 3.1 models the necessary capabilities that requires performing impersonation, man-in-the-middle, and replay attacks. The goals of this adversary are as follows: (Goal 1) Convincing either an *SN* or a *UN* to misbelieve that a legitimate partner is participating in a communication within the timeout period; (Goal 2) Having better strategy than the wild guess in distinguishing a session key k_s against a random string with the same length. We show that there is no adversary to effectively, and with non-negligible probability, achieve either of these goals.

Goal 1 happens when either *UN* accepts or *SN* accepts. We separately discuss these cases.

- The *UN* accepts. This happens if and only if $tid_N = h(id_N^*, t_N, c_N^*, r_N^*)$. We assume that the *SN* does not generate a tid_N after $t^* - \Delta T$, otherwise it violates definition of Goal 1. If this equation is true but the hash $h(id_N^*, t_N, c_N^*, r_N^*)$ has never been computed, this will happen only with $p = 2^{-\text{length}(h)}$.

If this equation is true and the hash has been computed before, we may conclude that it is not produced by a legitimate *SN* and *UN*. This is due to the fact that id_N is unique and *SN* does not produce any at t_N and *UN* would never send computed tid_N. Therefore, the only possibility is that the adversary computes the hash by itself. This happens only if the adversary has id_N and c_N which are not sent over the network. This is bounded by $p^2 \times q_h$ where q_h is the maximum number of the hashes that are able to query with reasonable resources.

- The *SN* accepts. This happens if and only if the value of the β is equal to $h(x_N^+, r_N, f_N^*, \eta, \mu)$. Similarly, if the hash was never computed, the probability is bounded by p. If the hash is previously computed by the *UN*, the same *SN* (with id_N^*) has already sent a login request with r_N^*. Since r_N^* is randomly chosen, this happens only with $p \times q_E$, where q_E is the total number of the sessions executed by the *SN*. Otherwise, the adversary should correctly guess id_N^* and c_N, which happen only with $p^2 \times q_h$.

To sum up, the occurrence of Goal 1 has a probability lower than $(q_E + 2)p + 2q_h p^2$, where $p = 2^{-\text{length}(h)}$, q_E is the total number of the sessions executed by the *SN*, and q_h is the total number of the hashes that are able to be computed by the adversary with reasonable resources. This number is negligible when the length of the hash is large.

Goal 2 happens only when the *UN* accepts and the hash $h(id_N, r_N, f_N, x_N)$ has been computed by the adversary since k_s is never transmitted. However, id_N and x_N are both secret. A correct guess of this variable is bounded by $p^2 \times q_h$.

Considering the probability to concurrently achieve the both Goals 1 and 2, an attacker may cast as an impersonation attack, a man-in-the-middle attack, or a replay attack has a probability less than $(q_E + 2)p + 3q_h p^2$.

5.3. Tracking Attacks and Anonymity

We may see that the tracking attack, mentioned in Section 3.3, no longer operates. First of all, an *FN* serves only as a relay to replay a message. No information can be harvested to identify the relay *FN*. Furthermore, the equality $a_2 \oplus b_2 = \eta_1 \oplus \mu_1$ no longer holds, where $\eta_1 \oplus \mu_1 = a_2 \oplus b_2 \oplus h(f_N, c_N) \oplus h(c_N, f_N)$. Since c_N and f_N are not computable by the adversary, computing $h(f_N, c_N)$ or $h(c_N, f_N)$ is not possible.

6. Simulation Verification Using a Proverif Tool

Proverif is an automatic cryptographic protocol verifier, which is widely used to specify and analyze the security of authenticated key agreement protocols [19–23].

In this section, we utilize Proverif to further analyze the security and validity of the proposed protocol. In this simulation, two main roles, *SN* and *UN*, are included. The whole simulation contains the following procedures:

- First, we need to define some variables used in this simulation. K_{UN} is the secret key H_N, and SK_{SN} and SK_{UN} are the final shared key established by *SN* and *UN*, respectively—then comes the functions and events (Figure 5),
- Second, we list the goals of this simulation. More specifically, our goals is to ensure that the whole authentication process is successful, the shared key can be established, and the attacker cannot obtain the key anyway (Figure 6),
- The process of *SN* (Figure 7),
- The process of *UN* (Figure 8),
- The main execution (Figure 9).
- According to the simulation results depicted in Figure 10, we can observe that the proposed protocol can achieve the goals mentioned in Figure 6.

```
(* channel*)
free ch:channel. (* public channel *)

(* shared keys *)
free SK_SN:bitstring [private].
free SK_UN:bitstring [private].

(* constants *)
free k_UN:bitstring [private]. (* the UN's secret key *)

(* functions & reductions & equations *)
fun h(bitstring):bitstring. (* hash function*)
fun senc(bitstring,bitstring):bitstring. (* symmetric encryption *)
reduc forall m:bitstring, key:bitstring; sdec(senc(m,key),key)=m.
fun con(bitstring,bitstring):bitstring. (* concatenation operation *)
fun xor(bitstring,bitstring):bitstring. (* XOR operation*)
equation forall m:bitstring, n:bitstring;xor(xor(m,n),n)=m.

(* event *)
event BeginAuth(bitstring).
event EndAuth(bitstring).
```

Figure 5. Proverif code of variables, functions and events.

```
(* queries *)
query attacker(SK_SN).
query attacker(SK_UN).
query id:bitstring; inj-event(BeginAuth(id)) ==> inj-event(EndAuth(id)).
```

Figure 6. Goal of this simulation.

```
(* SN's process *)
let processSN(idN:bitstring,aN:bitstring,bN:bitstring,cN:bitstring) =
  new rN:bitstring;
  new tN:bitstring;
  let xN = xor(idN,aN) in
  let yN = xor(rN,xN) in
  let tidN = h(con(xor(xor(idN,tN),cN),rN)) in
  out(ch,(tidN,yN,aN,bN,tN));

  in(ch,(alpha:bitstring,beta:bitstring,eta:bitstring,mu:bitstring));
  let fN_star = xor(alpha,cN) in
  let beta_star = h( con(con(con(con(idN,rN),fN_star),eta),mu) ) in
  if beta_star = beta then
  let aN_plus = xor(eta,h(con(fN_star,cN))) in
  let bN_plus = xor(mu,h(con(cN,fN_star))) in
  let ks_star = h( con(con(con(idN,rN),fN_star),xN) ) in

  out(ch,senc(SK_SN,ks_star));
  event EndAuth(idN).
```

Figure 7. Proverif code of *SN*.

```
(* UN's process *)
let processUN =
  in(ch,(tidN:bitstring,yN:bitstring,aN:bitstring,bN:bitstring,tN:bitstring));

  let kN_star = xor(xor(bN,aN),k_UN) in
  let xN_star = h(con(k_UN,kN_star)) in
  let idN_star = xor(xN_star,aN) in
  let rN_star = xor(yN,xN_star) in
  let cN_star = h(con(idN_star,k_UN)) in
  let tidN_star = h( con(xor(xor(idN_star,tN),cN_star),rN_star) ) in
  if tidN_star = tidN then
  (
    event BeginAuth(idN_star);
    new fN:bitstring;
    new kN_plus:bitstring;
    let aN_plus = xor(h(con(k_UN,kN_plus)),idN_star) in
    let bN_plus = xor(xor(kN_plus,k_UN),aN_plus) in
    let eta = xor(h(con(fN,cN_star)),aN_plus) in
    let mu = xor(h(con(cN_star,fN)),bN_plus) in
    let alpha = xor(cN_star,fN) in
    let beta = h( con(con(con(con(idN_star,rN_star),fN),eta),mu) )   in
    out(ch,(alpha,beta,eta,mu));

    let ks = h( con(con(con(idN_star,rN_star),fN),xN_star) ) in
    out(ch,senc(SK_UN,ks))
).
```

Figure 8. Proverif code of *HN*.

```
(* ----- Main ----- *)
process
  (* register a new sensor node *)
  new idN:bitstring;
  new kN:bitstring;
  let aN = xor(h(con(k_UN,kN)),idN) in
  let bN = xor(xor(kN,k_UN),aN) in
  let cN = h(con(idN,k_UN)) in
  ( !processSN(idN,aN,bN,cN) ) | ( !processUN )
```

Figure 9. Main process of this simulation.

```
-- Query inj-event(BeginAuth(id)) ==> inj-event(EndAuth(id))
RESULT inj-event(BeginAuth(id)) ==> inj-event(EndAuth(id)) is true.

-- Query not attacker(SK_UN[])
RESULT not attacker(SK_UN[]) is true.

-- Query not attacker(SK_SN[])
RESULT not attacker(SK_SN[]) is true.
```

Figure 10. Simulation results.

7. Performance Evaluation

This section describes performance evaluation of the repaired protocol along with other related protocols [4,10–12,14] in security properties and estimated time. We focus on the security against the anonymity, tracking attack, insider attack, replay attack, impersonation attack, man-in-the-middle attack, mutual authentication and the session key forward secrecy. From Table 1, we see that only the repaired protocol, Wu's protocol [12] and Shen et al. [14] fulfill all the security properties.

Table 1. Comparison of the security properties. Y and N stands for fulfilling and not fulfilling the requirement respectively.

	C1	C2	C3	C4	C5	C6	C7	C8
[10]	Y	Y	N	Y	Y	Y	Y	Y
[11]	N	N	Y	Y	Y	Y	Y	Y
[12]	Y	Y	Y	Y	Y	Y	Y	Y
[4]	N	N	N	Y	N	Y	Y	Y
[14]	Y	Y	Y	Y	Y	Y	Y	Y
Ours	Y	Y	Y	Y	Y	Y	Y	Y

C1: Provide anonymity;
C2: Withstand tracking attack;
C3: Withstand insider attack;
C4: Withstand repay attack;
C5: Withstand impersonation attack;
C6: Withstand man-in-the-middle attack;
C7: Mutual authentication;
C8: The session key forward secrecy

We analyze the time performance of these protocol by analysis of the core cryptographic operations used in each of them, and then estimate the running time of these protocols by adding the time of executed cryptographic operations. We do not consider the possibility of parallel computation with multi-core technologies since most wearable devices are only single core. Pipelining is also not discussed here since the authentication usually needs to be executed once.

We consider two possible realizations of an *SN*. A sensor device using the MICAz with 4 KB RAM (Crossbow Technology, San Jose, CA, USA) and 7-MHz ATmega128L microcontroller (Microchip Technology Inc, Chandler, AZ, USA) and a smart phone using an iPhone 6s (Apple, Cupertino, CA, USA) with 2 GB RAM ARM (armv8-a) CPU. The data are taken from [13,24,25] for the time required on the MICAz while we implement those implementations on a smart phone using the Pairing Based Cryptographic Library [26]. The result is summarized in Table 2.

Table 2. Computation of the cryptographic operations.

Symbol	Description	Running Time on a Smartphone	Running Time on a MICAz
T_h	Hash function	0.03 ms	8 ms [25]
T_{sym}	Symmetric encryption/description operation	0.12 ms	3.5 ms [24]
T_{sm}	Scalar multiplication over elliptic curves	20.23 ms	2450 ms [13]
T_{bp}	Bilinear pairing operation	25.64 ms	5320 ms [13]

Table 3 lists the estimated time of the mentioned protocols, considering the above experimental data. From this table, we may observe that the repaired protocol costs more time than Li's protocol [4] as it takes six more hash functions, but costs less time than the other related protocols [10–12,14] .

Table 3. Comparison of the estimated time.

Protocols	Time Cost	Running Time on a Smartphone	Running Time on a MICAz
[10]	$4T_h + 5T_{sm} + 3T_{bp}$	178.19 ms	28242 ms
[11]	$11T_h + 9T_{sm} + 3T_{sym}$	182.64 ms	22148.5 ms
[12]	$7T_h + 8T_{sm} + T_{bp} + 2T_{sym}$	187.93 ms	24983 ms
[4]	$9T_h$	0.27 ms	72 ms
[14]	$9T_h + 13T_{sm}$	263.26 ms	31922 ms
Ours	$15T_h$	0.45 ms	120 ms

8. Conclusions

We demonstrated that Li's protocol is broken and should not be used in any application implementation related to the WBAN. At the same time, we proposed another architecture that research should be considered when designing any authentication. In this architecture, the linear relationship connecting an *SN* to an *FN* and an *FN* to an *HN* is abandoned. Instead, *SN* s, *FN* s and *HN* s are directly connected to each other through a pairwise secret. The *FN* changes its role in an *SN-HN* communication from coordinating to relaying messages between the *SN* and *HN*. We believe that this approach is highly effective and secure so that compromise of the *HN* or *FN* would not lead to a total compromise of the system. In such an architecture, an *FN* may be abused through consuming the relay service by attackers. This problem, however, appears in most of the relaying systems in all wireless networks, which may be handled via some firewall rules or intrusion detection techniques. This represents an interesting research topic to be further studied by the authors in the future.

Author Contributions: C.-M.C. and K.-H.W. wrote the main concepts of the manuscript; B.X. designed and implemented the experiments; T.-Y.W. checked the English writing and organization of the manuscript.

Funding: The work of Chien-Ming Chen was supported in part by Shenzhen Technical Project under Grant number JCYJ20170307151750788 and in part by Shenzhen Technical Project under Grant number KQJSCX20170327161755. The work of Tsu-Yang Wu was supported in part by the Science and Technology Development Center, Ministry of Education, China under Grant no. 2017A13025 and the Natural Science Foundation of Fujian Province under Grant no. 2018J01636.

References

1. Dussault, C.; Toeg, H.; Nathan, M.; Wang, Z.J.; Roux, J.F.; Secemsky, E. Electrocardiographic Monitoring for Detecting Atrial Fibrillation After Ischemic Stroke or Transient Ischemic Attack. *Circ. Arrhythm. Electrophysiol.* **2015**, *8*, 263–269. [CrossRef] [PubMed]

2. Epstein, L.J.; Kristo, D.; Strollo, P.J.; Friedman, N.; Malhotra, A.; Patil, S.P.; Ramar, K.; Rogers, R.; Schwab, R.J.; Weaver, E.M.; et al. Clinical guideline for the evaluation, management and long-term care of obstructive sleep apnea in adults. *J. Clin. Sleep Med.* **2009**, *5*, 263–276. [PubMed]

3. Toorani, M. On Vulnerabilities of the Security Association in the IEEE 802.15.6 Standard. In Proceedings of the Financial Cryptography and Data Security: FC 2015 International Workshops, BITCOIN, WAHC, and Wearable, San Juan, Puerto Rico, 26–30 January 2015; pp. 245–260.

4. Li, X.; Ibrahim, M.H.; Kumari, S.; Sangaiah, A.K.; Gupta, V.; Choo, K.K.R. Anonymous mutual authentication and key agreement scheme for wearable sensors in wireless body area networks. *Comput. Netw.* **2017**, *129*, 429–443. [CrossRef]

5. Kaufman, C.; Hoffman, P.; Nir, Y.; Eronen, P. *Internet Key Exchange Protocol Version 2 IKEv2*; RFC 5996, RFC Editor; IETF: Fremont, CA, USA, 2010.

6. Wang, K.H.; Chen, C.M.; Fang, W.; Wu, T.Y. On the security of a new ultra-lightweight authentication protocol in IoT environment for RFID tags. *J. Supercomput.* **2017**, *74*, 1–6. [CrossRef]

7. Wang, D.; Wang, P. On the anonymity of two-factor authentication schemes for wireless sensor networks: Attacks, principle and solutions. *Comput. Netw.* **2014**, *73*, 41–57. [CrossRef]

8.   Keoh, S.L.; Lupu, E.; Sloman, M. Securing body sensor networks: Sensor association and key management. In Proceedings of the 2009 IEEE International Conference on Pervasive Computing and Communications, PerCom 2009, Galveston, TX, USA, 9–13 March 2009; pp. 1–6.

9.   Liu, J.; Kwak, K.S. Hybrid security mechanisms for wireless body area networks. In Proceedings of the 2010 Second International Conference on Ubiquitous and Future Networks (ICUFN), Jeju, Korea, 16–18 June 2010; pp. 98–103.

10. Liu, J.; Zhang, Z.; Chen, X.; Kwak, K.S. Certificateless remote anonymous authentication schemes for wirelessbody area networks. *IEEE Trans. Parallel Distrib. Syst.* **2014**, *25*, 332–342. [CrossRef]

11. Zhao, Z. An efficient anonymous authentication scheme for wireless body area networks using elliptic curve cryptosystem. *J. Med. Syst.* **2014**, *38*, 13. [CrossRef] [PubMed]

12. Wu, L.; Zhang, Y.; Li, L.; Shen, J. Efficient and anonymous authentication scheme for wireless body area networks. *J. Med. Syst.* **2016**, *40*, 134. [CrossRef] [PubMed]

13. Xiong, X.; Wong, D.S.; Deng, X. TinyPairing: A Fast and Lightweight Pairing-Based Cryptographic Library for Wireless Sensor Networks. In Proceedings of the 2010 IEEE Wireless Communication and Networking Conference, Sydney, NSW, Australia, 18–21 April 2010; pp. 1–6.

14. Shen, J.; Gui, Z.; Ji, S.; Shen, J.; Tan, H.; Tang, Y. Cloud-aided lightweight certificateless authentication protocol with anonymity for wireless body area networks. *J. Netw. Comput. Appl.* **2018**, *106*, 117–123. [CrossRef]

15. Venkatasubramanian, K.K.; Banerjee, A.; Gupta, S.K.S. PSKA: Usable and secure key agreement scheme for body area networks. *IEEE Trans. Inf. Technol. Biomed.* **2010**, *14*, 60–68. [CrossRef] [PubMed]

16. Zhang, Z.; Wang, H.; Vasilakos, A.V.; Fang, H. ECG-cryptography and authentication in body area networks. *IEEE Trans. Inf. Technol. Biomed.* **2012**, *16*, 1070–1078. [CrossRef] [PubMed]

17. Shi, L.; Yuan, J.; Yu, S.; Li, M. ASK-BAN: Authenticated secret key extraction utilizing channel characteristics for body area networks. In Proceedings of the Sixth ACM Conference on Security and Privacy in Wireless and Mobile Networks, Budapest, Hungary, 17–19 April 2013; ACM: New York, NY, USA, 2013; pp. 155–166.

18. Wang, K.H.; Chen, C.M.; Fang, W.; Wu, T.Y. A secure authentication scheme for Internet of Things. *Pervasive Mob. Comput.* **2017**, *42*, 15–26. [CrossRef]

19. Jiang, Q.; Zeadally, S.; Ma, J.; He, D. Lightweight three-factor authentication and key agreement protocol for internet-integrated wireless sensor networks. *IEEE Access* **2017**, *5*, 3376–3392. [CrossRef]

20. Chaudhry, S.A.; Naqvi, H.; Sher, M.; Farash, M.S.; Hassan, M.U. An improved and provably secure privacy preserving authentication protocol for SIP. *Peer-to-Peer Netw. Appl.* **2017**, *10*, 1–15. [CrossRef]

21. Wu, F.; Xu, L.; Kumari, S.; Li, X.; Shen, J.; Choo, K.K.R.; Wazid, M.; Das, A.K. An efficient authentication and key agreement scheme for multi-gateway wireless sensor networks in IoT deployment. *J. Netw. Comput. Appl.* **2017**, *89*, 72–85. [CrossRef]

22. Abbasinezhad-Mood, D.; Nikooghadam, M. Efficient anonymous password-authenticated key exchange protocol to read isolated smart meters by utilization of extended chebyshev chaotic maps. *IEEE Trans. Ind. Inform.* **2018**. [CrossRef]

23. Abbasinezhad-Mood, D.; Nikooghadam, M. Design and hardware implementation of a security-enhanced elliptic curve cryptography based lightweight authentication scheme for smart grid communications. *Future Gener. Comput. Syst.* **2018**, *84*, 47–57. [CrossRef]

24. Panait, C.; Dragomir, D. Measuring the performance and energy consumption of AES in wireless sensor networks. In Proceedings of the 2015 Federated Conference on Computer Science and Information Systems (FedCSIS), Lodz, Poland, 13–16 September 2015; pp. 1261–1266.

25. Koschuch, M.; Hudler, M.; Saffer, Z. Towards algorithm agility for wireless sensor networks: Comparison of the portability of selected Hash functions. In Proceedings of the 2013 International Conference on Data Communication Networking (DCNET), Reykjavik, Iceland, 29–31 July 2013; pp. 1–5.

26. Lynn, B. On the Implementation of Pairing-Based Cryptosystems. Ph.D. Thesis, Stanford University Stanford, Stanford, CA, USA, 2007.

Article

Respiration Symptoms Monitoring in Body Area Networks

Lu Liu [1,2,†]**, Syed Aziz Shah** [1,†]**, Guoqing Zhao** [1] **and Xiaodong Yang** [1,*]

[1] School of Electronic Engineering, Xidian University, Xi'an 710071, China; liulu88131@163.com (L.L.); azizshahics@yahoo.com (S.A.S.); 15091775293@163.com (G.Z.)
[2] School of Communications and Information Engineering, Xi'an University of Posts and Telecommunications, Xi'an 710061, China
* Correspondence: xdyang@xidian.edu.cn
† These authors contributed equally to the work.

Received: 8 March 2018; Accepted: 1 April 2018; Published: 6 April 2018

Abstract: This work presents a framework that monitors particular symptoms such as respiratory conditions (abnormal breathing pattern) experienced by hyperthyreosis, sleep apnea, and sudden infant death syndrome (SIDS) patients. The proposed framework detects and monitors respiratory condition using S-Band sensing technique that leverages the wireless devices such as antenna, card, omni-directional antenna operating in 2 GHz to 4 GHz frequency range, and wireless channel information extraction tool. The rhythmic patterns extracted using S-Band sensing present the periodic and non-periodic waveforms that correspond to normal and abnormal respiratory conditions, respectively. The fine-grained amplitude information obtained using aforementioned devices is used to examine the breathing pattern over a period of time and accurately identifies the particular condition.

Keywords: S-Band sensing; wireless channel information; hyperthyreosis/sleep apnea disease monitoring

---

1. Introduction

Hyperthyreosis is a condition that is characterized by the excess production of thyroid hormone [1]. This disorder is primarily caused by the auto-immune system, produces a butterfly-shaped organ in the neck and controls the rate of metabolism. The term hyperthyroidism refers to the production of the huge amount of thyroid hormone that affects the metabolism rate, as a result increase in shortness of breath (abnormal breathing), mood swings, anxiety, muscle weakness, hair loss and tremors in limbs can be experienced [2]. Sleep apnea is a sleep disorder that causes abnormal breathing which can ultimately cause apnea [3], while sudden infant death syndrome condition is experienced by a newly born baby or around one year old where the baby experience abnormal respiration and then suddenly stops breathing while asleep. The normal breathing rate for healthy person while is resting is 12 to 20 breaths per minute and the abnormal respiratory rate is over 25 breaths per minute. Normal breathing presents a periodic waveform and abnormal breathing waveform is non-period.

In healthcare environment, the introduction of wireless communication systems has promoted the improvement in the efficiency of patient care and health management [4,5]. This paper presents S-Band sensing technique (operating at 2 GHz to 4 GHz) that leverage wireless channel information (WCI) obtained using wireless devices such as antenna, card, omni-directional antenna, etc. to monitor and detect the respiratory condition experienced by patient suffering from diseases such as hyperthyreosis, sleep apnea and sudden infant death syndrome (SIDS). The aim is to assist the doctors or caregivers for documenting objective assessments.

2. Related Work

This section presents the research work performed for detecting the respiratory patterns. With regard to breathing pattern assessment, polysomnography (PSG) used pressure transducers in the nostrils to record the airflow pressure [6]. The subject has to sleep while wearing multiple sensors including an Electroencephalography (EEG) monitor, an Electromyography (EMG) monitor, an Electrooculography (EOG) monitor, nasal probes, etc. [7]. The system introduced in [8] exploits the pressure sensor array equipped with a mat to detect the subject's breathing pattern, and body motion. The camera-based system introduced in [9] monitors the patient and extract the respiration rate. However, the sleeping mat equipped with sensors is expensive, uncomfortable at times and the vision-based system is light dependent, computationally expensive, and raises privacy issues.

As an alternative, radio frequency (RF) signals can be used to track the subject's chest movements due to the advantage of less complexity and independence of light intensity. The core idea is to exploit the change in wireless signal propagation between transceiver pair due to minute chest movement. Leveraging RF technologies, Aqib et al. [10] used the Frequency Modulated Carrier Waves (FMCW) radar to detect subject's breathing pattern. The main limitation of FMCW radar is that the transmitted signals interfere with the nearby RF systems since it exploits large range of frequencies.

Patwari et al. [11] used the commercially available small wireless devices to track breathing pattern and exploited the received signal strength indicator (RSSI). Further improvements were brought up by [12] where the RSSI measurements were obtained using single transceiver pair to track the subject's respiration. The RSSI data, however, do not report the subtle chest movement due to breathing, as a result, the desirable results can be greatly affected by environmental noise. In order to cope with this issue, the RSSI measurements are essentially termed as sinusoidal signals, and the amplitude information is generally estimated on the basis receive signal measurements under particular periodic assumption [13]. Also, the performance of RSSI dramatically degrades in complex situations due to temporal dynamics and multipath fading [14]. However, the RSSI-based systems give us a blueprint to use wireless channel features that are more sensitive than RSSI, in extracting the normal breathing pattern and abnormal breathing pattern of a patient suffering from hyperthyreosis, sleep apnea and SIDS. Thus, in order to extract the subject's normal and abnormal respiration, we use S-Band sensing technique which is highly sensitive to body motion by exploiting the wireless channel information.

3. S-Band Sensing and Breath Detection

The wireless channel information characterizes the wireless channel properties of a communication link. The information essentially presents the RF signal propagation from transmitter to receiver and describes the combined effect of multipath propagation such as reflection, fading, and scattering [15]. The signal received can be expressed as:

$$\mathbf{H}(i) = \left\| \mathbf{H}(i) \right\| e^{(j\angle \mathbf{H}(i))}. \tag{1}$$

Here $\mathbf{H}(i)$ is the wireless channel information, the terms $\|\mathbf{H}(i)\|$ and $\angle \mathbf{H}(i)$ denote the amplitude and phase information for ith frequency, respectively. It should be noted that the card reveals group of 30 frequencies known as one WCI packet and is intrinsic feature of the device that sweeps over multiple channel frequencies.

The experimental design was implemented in a meeting room at Xidian University, as shown in Figure 1. An antenna which is in the range of S-Band (2 GHz to 4 GHz) was deployed as the transmitter and a 2.4 GHz omni-directional antenna working as a receiver. The wireless channel information extraction tool introduced by [16] was installed. The host computer receive 10 WCI packets per second [17]. From each WCI packet received using the specific NIC, a *30 x 1* matrix was extracted known as channel frequency response (CFR) and can be written as:

$$CFR_{matrix} = \left[h^1, h^2, h^3, \ldots, h^n \right]. \tag{2}$$

Here h^i is a complex number and describes channel frequency response of the *i*th frequency and $n = 30$ is the total number of frequencies. In order to examine the time history of a total number of WCI packets received over a period of time, the CFR_{matrix} is put together recorded at various time intervals and forms a WCI time history stream as:

$$CFR_{time\_history} = [CFR_{matrix}^{1}, CFR_{matrix}^{2}, CFR_{matrix}^{3}, \ldots CFR_{matrix}^{k}]. \tag{3}$$

Here $CFR_{time\_history}$ is a *30-x-1* matrix, where k represents the total number of WCI packets received within a particular time window.

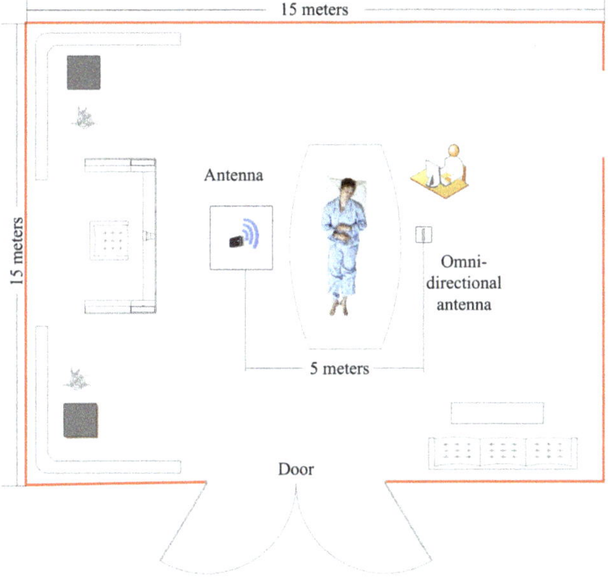

Figure 1. Experimental setting for respiration monitoring.

Figure 2 indicates the variances of raw WCI due to the subject's chest movement, which will be used to extract the breathing pattern in the subsequent section. In order to extract the breathing pattern, we select the frequency # 13 as in [15]. We have examined all the 30 frequencies for detecting breathing pattern and compared the results with digital respiratory sensor. The main advantage of using S-Band sensing which exploit the NIC is that one or multiple frequencies out of 30 can be used for desired purpose. We tested the proposed framework in 8 various places that had different geometry and structure. In 7 cases, frequency # 13 provided the best results when observed through the digital respiratory sensor. Only at one place, the frequency # 12 provided results closely matched to the digital respiratory sensor. The primary reason that frequency # 13 provides best results because it is available to all wireless local area network systems but rarely used making it less prone to interference.

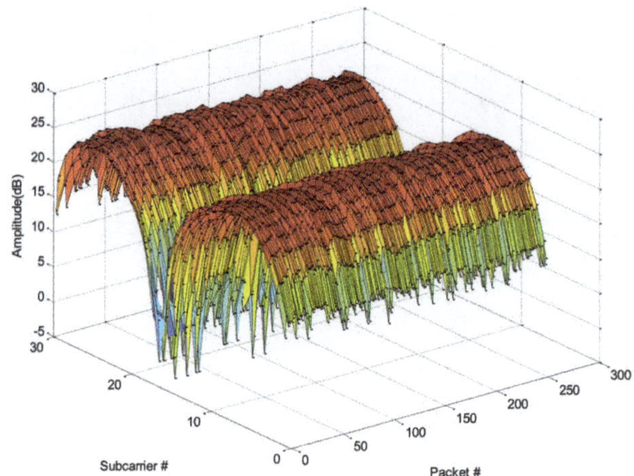

Figure 2. The raw wireless channel information (WCI) amplitude information received for breathing pattern detection.

4. Rational of Using S-Band Sensing for Breath Detection

This section gives an intuition about how the S-Band sensing technique leveraging wireless channel information can be used to track subtle chest movements for breath detection. Ideally, an RF signal propagates directly from transmitter to receiver when no obstacle is encountered, is known as line-of-sight (LOS) communication. When an obstacle is encountered, the signal experiences scattering, reflection, and diffraction. This is known as non-line-of-sight (NLOS) communication. The signal received through LOS and NLOS communication will have different WCI data [18,19]. The signal propagation continuously moves back and forth between LOS and NLOS due to person's chest movement, as a result the WCI data retrieved are significantly different. For breathing pattern detection, in almost all scenarios, the minute chest movement keeps on changing the communication path between LOS and NLOS. The RF signal propagating from transmitter to the receiver will always have a LOS communication path and a lot of NLOS communication paths. When the subject is inhaling and exhaling, the received RF signal will continuously keep on changing due to reflections caused by the person's chest movement.

5. Overview of the Proposed Method

The system architecture designed to monitor the breathing of patient suffering from hyperthyreosis, sleep apnea and SIDS diseases consist of two main components as shown in Figure 3. Identification of breathing pattern using a wearable digital respiratory sensor, and breathing pattern monitoring using S-Band sensing technique leveraging wireless channel information. From raw WCI data, the amplitude information is extracted which is examined against the time history using an individual frequency. The variances of amplitude information are then passed through a filtering process, using wavelet filter. The variances of amplitude information are then analyzed and compared with the data collected with a wearable sensor to see if they can be attributed to the normal breathing or abnormal breathing.

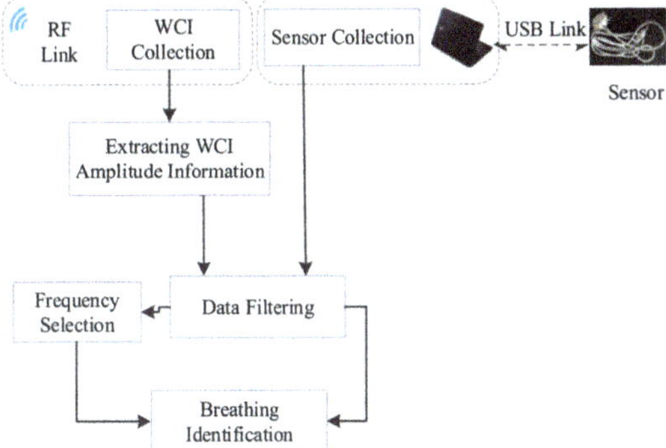

Figure 3. System architecture for breath monitoring.

Noise Filtering Process

The noise contained in the data collected using the wearable sensor as well as in the wireless channel information should be removed. The wavelet filter presented in [20] is used to remove the noise and smoothens the rising and falling edges appeared in the received signals critical for tracking the normal and abnormal breathing. The advantage of using wavelet filter over other filters such as Chebyshev or Butterworth is due to its better performance considering the high-frequency noise.

6. Tracking Person's Breathing in Typical Environment

This section presents the details regarding the proposed method of tracking breathing under typical conditions when person lay on the back. The individual CFR sequence i.e., the individual frequency is selected to accurately estimate the breathing pattern of a person. Figure 4 shows the variations of amplitude information of all 30 frequencies against packet number retrieved over a period of time.

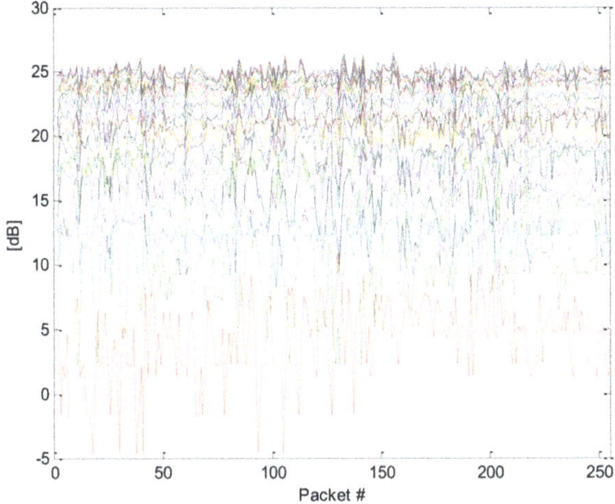

Figure 4. The channel frequency response (CFR) sequence of all frequencies obtained during experiment.

6.1. Analysis of Measured Data for Normal Breathing

Figure 5 indicates the breathing pattern of a person lying in a straight position on the back. The measurements were taken for 60 s. Figure 5a shows the raw breathing pattern obtained using respiratory sensor and Figure 5b shows the filtered estimated breathing pattern. The person, in this case, completed nearly 16 breathing cycles. We further examine the breathing pattern obtained using S-Band sensing leveraging the wireless channel information.

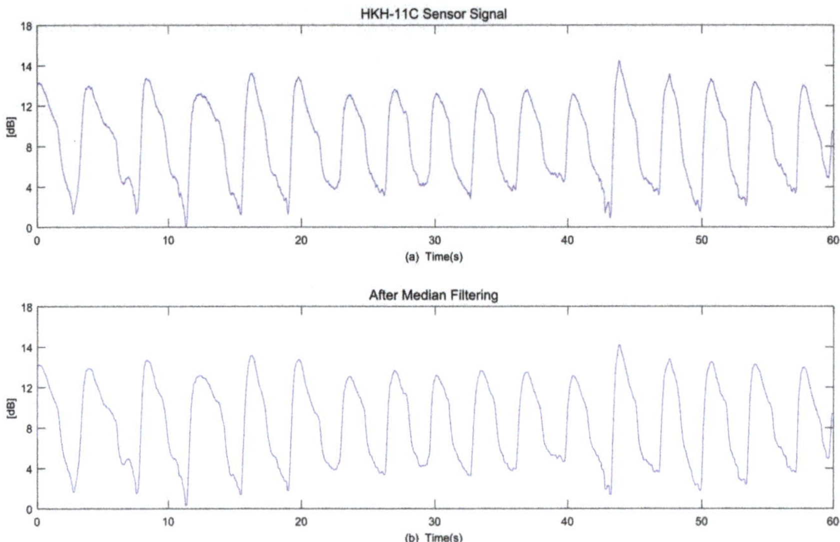

Figure 5. Estimated breathing pattern obtained using Digital Respiratory Sensor. (**a**) Raw breathing pattern obtained; (**b**) Filtered breathing pattern obtained.

The frequency-selective fading poses various extents on the RF signal considering different frequencies. As shown in Figure 6, different frequencies have different variations with respect to time for different power levels. Thus we select frequency # 13 that best describes the breathing pattern of the subject as illustrated in Figure 6a. The data obtained for the particular frequency was then passed into the wavelet filter to remove the sharp edges and smooth the signal. Comparing the breathing pattern shown in Figure 6b with the one in Figure 5b, we observe that both the waves forms obtained are identical considering the number of cycles during the same time period.

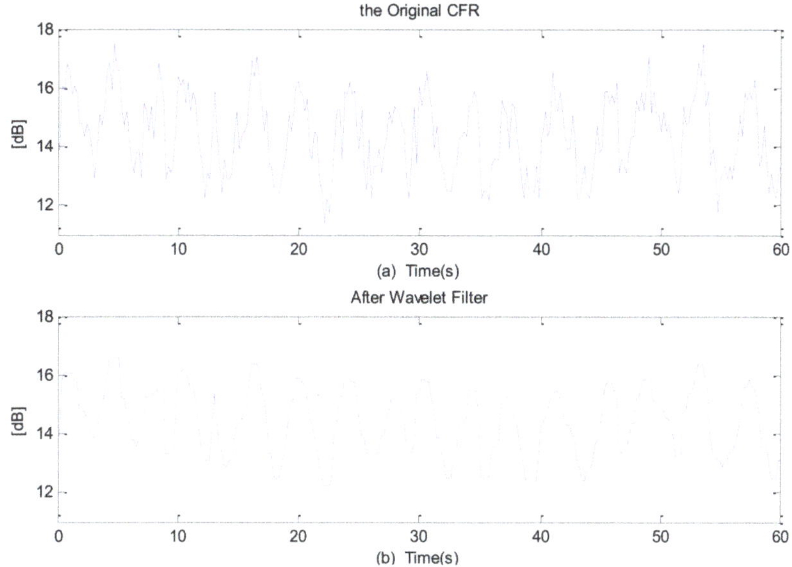

Figure 6. The breathing pattern obtained using S-Band sensing. (**a**) Variances of raw amplitude information of frequency # 13; (**b**) Filtered breathing pattern of the individual frequency.

6.2. Tracking Abnormal Breathing

In the prevision section, we described how S-Band sensing technique can be used to track person's normal breathing and compared the measured data with that of a wearable sensor. We now present the measurements obtained for tracking person's abnormal breathing, one of the conditions experienced by a patient suffering from hyperthyreosis disease. We used the typical experimental environment and asked the volunteer to breathe abnormally. Hence, we collected the raw amplitude information for 30 frequencies by retrieving 200 WCI packets over a period of time as shown in Figure 7.

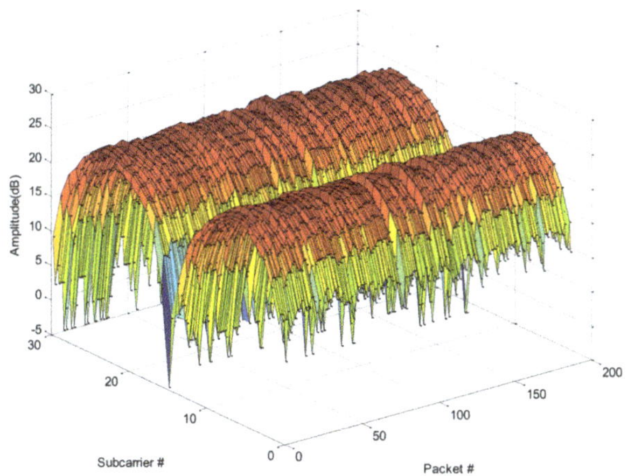

Figure 7. Raw amplitude information received for tracking abnormal breathing.

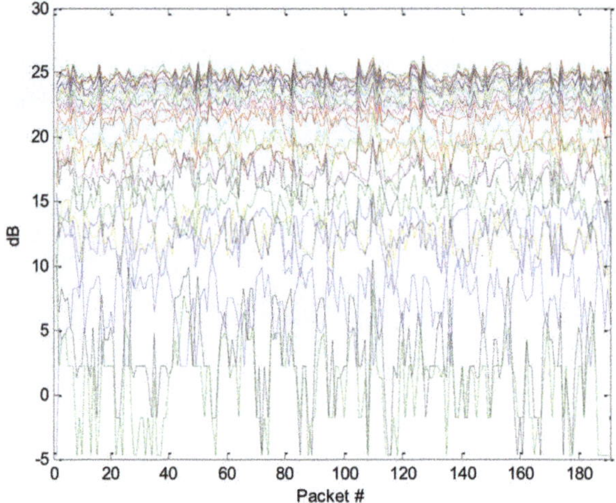

Figure 8. The CFR sequence of all frequencies obtained for abnormal breathing.

We follow the same procedure as described earlier, and examine all the frequencies to extract the abnormal breathing. The patient was observing abnormal breathing during the process and we opted the frequency # 13 for data analysis from Figure 8. Thus we examine the same frequency for tracking abnormal breathing and compare the measured results with wearable breathing sensor.

Figure 9 shows the data obtained using the wearable sensor to track the person's abnormal breathing pattern. The raw data measured for 60 s show clear signs that the person was experiencing abnormal breathing while filtering out the measured data also indicate the same. A clearly non-periodicity can be observed. We further analyze the data retrieved using S-Band sensing as shown in Figure 10a that present raw breathing patterns and Figure 10b which shows filtered breathing pattern and see whether it is able to detect the abnormal sleeping pattern or not.

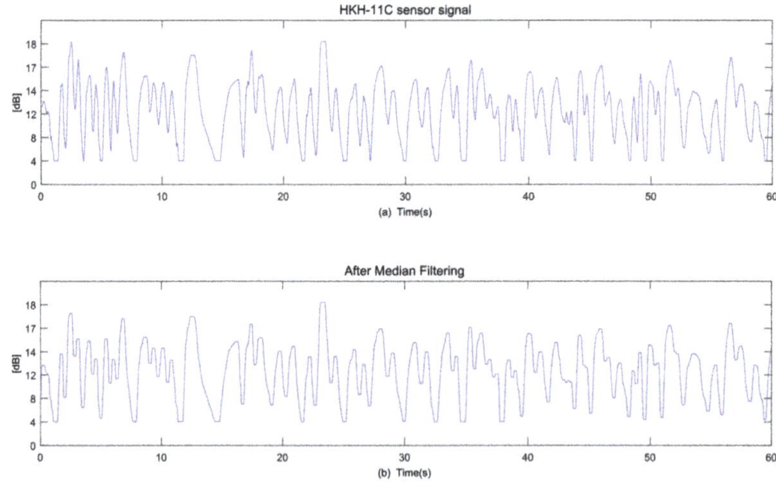

Figure 9. Abnormal breathing pattern tracked using Digital Respiratory Sensor. (**a**) The original data measured for abnormal breathing; (**b**) Filtered abnormal breathing pattern obtained.

The person in this scenario was asked to breathe abnormally which was finely tracked by the wearable sensor. The primary indicator in the particular case was the breathing sensor which did not present any periodicity. The same principle is applied here. The raw and filtered amplitude information indicates a clear non-periodicity for a period of time which indicates the abnormal breathing when the subject was laying on the back (straight position).

Spearman's Rank Correlation

We used Spearman's rank correlation to find out correlation between the breathing signals obtained through wearable sensor and S-Band. The Spearman correlation is based on rank method. The coefficient of correlation is calculated by rank differences. On the basis of rank differences, we find out the correlation between two signals. The advantage of Spearman's correlation coefficient is that the computation is quicker, especially when number of observations is limited. The method is useful for series of data such as rank, scores, etc. The formula for Spearman's correlation coefficient is as written follows:

$$r_k = 1 - \frac{6\sum D^2}{N^3 - N} \tag{4}$$

Here r_k is the coefficient of rank correlation, D denotes the rank differences, and N is the number of pair observations. The value of r_k lies between +1 and -1. Positive value shows that the direction of rank is the same and negative value shows that the ranks are given in opposite direction. We have compared the two signals obtained using wearable sensor and S-Band sensing as shown in Figures 5b and 6b, respectively. The Spearman's correlation coefficient obtained is 0.89 which employs that the value is positive and close to +1. We then compare the two waveforms for abnormal breathing as in Figures 9b and 10b. The r_k value obtained is 0.86 which is positive that means the direction of rank is same. Hence we can conclude that the signals obtained using wearable sensor and S-Band are identical considering both cases.

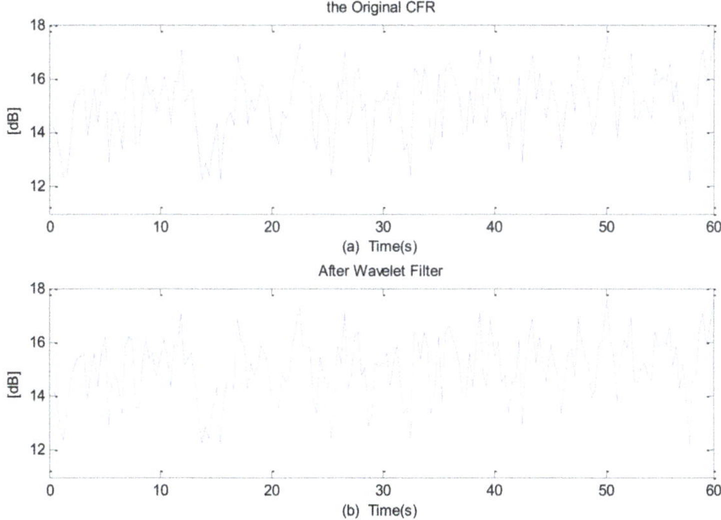

Figure 10. Abnormal breathing tracked using S-Band. (**a**) Variations of raw amplitude information for frequency # 13; (**b**) Variances of amplitude information filtered using wavelet filter.

7. Conclusions

This paper presented monitoring of vital signs experienced by patients suffering from various diseases such as hyperthyreosis, sleep apnea and SIDS diseases using S-Band sensing technique. The results obtained indicate that the wireless channel information can be used to monitor the breathing pattern and efficiently detects the abnormal respiratory when compared to the data retrieved using wearable sensor. However, the results we presented were preliminary findings of the breathing pattern of a person lying in supine position. The experiment was performed eight times on six different subjects at eight different locations lying in particular posture. The authors aim is to use the proposed framework as a foundation to develop an advanced level model that can detect subtle chest movement for various body postures on bed such as right lateral, left lateral, and fowler. The model would also be able to detect the patient's heartbeat which is one of the vital signs that needs to be monitored in order to assists the caregiver or doctor in home or hospital environment.

Acknowledgments: The authors would like to thank Dou Fan for data processing.

Author Contributions: Lu Liu, Syed Aziz Shah performed the experiment and wrote the paper; Guoqing Zhao and Xiaodong Yang contributed the idea.

Conflicts of Interest: The authors declare no conflict of interest.

References

1. Devereaux, D.; Tewelde, S.Z. Hyperthyroidism and thyrotoxicosis. *Emerg. Med. Clin. North Am.* **2014**, *32*, 277–292. [CrossRef] [PubMed]
2. De Leo, S.; Lee, S.Y.; Braverman, L.E. Hyperthyroidism. *Lancet* **2016**, *388*, 906–918. [CrossRef]
3. Yoon, H.N.; Choi, S.H.; Kwon, H.B.; Kim, S.K.; Hwang, S.H.; Oh, S.M.; Choi, J.W. Sleep-Dependent Directional Coupling of Cardiorespiratory System in Patients with Obstructive Sleep Apnea. *IEEE Trans. Biom. Eng.* **2018**, *PP*, 1. [CrossRef]
4. De Miguel-Bilbao, S.; Aguirre, E.; Iturri, P.L. Evaluation of Electromagnetic Interference and Exposure Assessment from s-Health Solutions Based on Wi-Fi Devices. *BioMed Res. Int.* **2015**, *2015*. [CrossRef] [PubMed]
5. Solanas, A.; Patsakis, C.; Conti, M.; Vlachos, I.S.; Ramos, V.; Falcone, F.; Martinez-Balleste, A. Smart health: A context-aware health paradigm within smart cities. *IEEE Commun. Mag.* **2014**, *52*, 74–81. [CrossRef]
6. Ancoli-Israel, S.; Chesson, A.; Quan, S.F. *The AASM Manual for the Scoring of Sleep and Associated Events: Rules, Terminology and Technical Specifications*; American Academy of Sleep Medicine: Darien, IL, USA, 2007.
7. Zhao, M.; Yue, S.; Katabi, D.; Jaakkola, T.S.; Bianchi, M.T. Learning Sleep Stages from Radio Signals: A Conditional Adversarial Architecture. In Proceedings of the 34th International Conference on Machine Learning, Sydney, Australia, 6–11 August 2017.
8. Li, W.; Sun, C.; Yuan, W.; Gu, W.; Cui, Z.; Chen, W. Smart Mat System with Pressure Sensor Array for Unobtrusive Sleep Monitoring. In Proceedings of the 39th Annual International Conference of the IEEE Engineering in Medicine and Biology Society (EMBC), Jeju Island, Korea, 11–15 July 2017; pp. 177–180.
9. Lim, S.H.; Golkar, E.; Rahni, A.A.A. Respiratory Motion Tracking Using the Kinect Camera. In Proceedings of the 2014 IEEE Conference on Biomedical Engineering and Sciences (IECBES), Sarawak, Malaysia, 8–10 December 2014; pp. 797–800.
10. Adib, F.; Kabelac, Z.; Mao, H.; Katabi, D.; Miller, R.C. Demo: Real-time Breath Monitoring Using Wireless Signals. In Proceedings of the 20th Annual International Conference on Mobile Computing and Networking, Maui, HI, USA, 7–11 September 2014; pp. 261–262.
11. Patwari, N.; Wilson, J.; Ananthanarayanan, S.; Kasera, S.K.; Westenskow, D.R. Monitoring breathing via signal strength in wireless networks. *IEEE Trans. Mob. Comput.* **2014**, *13*, 1774–1786. [CrossRef]
12. Kaltiokallio, O.J.; Yigitler, H.; Jäntti, R.; Patwari, N. Non-Invasive Respiration Rate Monitoring Using a Single COTS TX-RX Pair. In Proceedings of the 13th International Symposium on Information Processing in Sensor Networks, Berlin/Heidelberg, Germany, 15–17 April 2014; pp. 59–70.
13. Yang, X.; Shah, S.A.; Ren, A.; Fan, D.; Zhao, N.; Cao, D.; Tian, J. Detection of Essential Tremor at the S-Band. *IEEE J. Trans. Eng. Health Med.* **2018**, *6*, 1–7. [CrossRef] [PubMed]
14. Shah, S.A.; Zhao, N.; Ren, A.; Zhang, Z.; Yang, X.; Yang, J.; Zhao, W. Posture Recognition to Prevent Bedsores for Multiple Patients Using Leaking Coaxial Cable. *IEEE Access* **2016**, *4*, 8065–8072. [CrossRef]
15. Shah, S.A.; Ren, A.; Fan, D.; Zhang, Z.; Zhao, N.; Yang, X.; Luo, M.; Wang, W.; Hu, F.; Rehman, M.U.; et al. Internet of Things for Sensing: A Case Study in the Healthcare System. *Appl. Sci.* **2018**, *8*, 508. [CrossRef]

16. Halperin, D.; Hu, W.; Sheth, A.; Wetherall, D. Tool release: Gathering 802.11 n traces with channel state information. *ACM SIGCOMM Comput. Commun. Rev.* **2011**, *41*, 53. [CrossRef]

17. Yang, X.; Shah, S.A.; Ren, A.; Zhao, N.; Fan, D.; Hu, F.; Tian, J. Wandering Pattern Sensing at S-Band. *IEEE J. Biom. Health Inf.* **2017**. [CrossRef]

18. Shah, S.A.; Zhang, Z.; Ren, A.; Zhao, N.; Yang, X.; Zhao, W.; Hao, Y. Buried Object Sensing Considering Curved Pipeline. *IEEE Antenna. Wirel. Propag. Lett.* **2017**, *16*, 2771–2775. [CrossRef]

19. Dong, B.; Ren, A.; Shah, S.A.; Hu, F.; Zhao, N.; Yang, X.; Abbasi, Q.H. Monitoring of atopic dermatitis using leaky coaxial cable. *Healthc. Technol. Lett.* **2017**, *4*, 244–248. [CrossRef] [PubMed]

20. Villasenor, J.D.; Belzer, B.; Liao, J. Wavelet filter evaluation for image compression. *IEEE Trans. Image Proc.* **1995**, *4*, 1053–1060. [CrossRef] [PubMed]

Article

Internet of Things for Sensing: A Case Study in the Healthcare System

Syed Aziz Shah [1], Aifeng Ren [2], Dou Fan [2], Zhiya Zhang [2], Nan Zhao [2], Xiaodong Yang [2,*],
Ming Luo [2], Weigang Wang [3], Fangming Hu [2], Masood Ur Rehman [4], Osamah S. Badarneh [5] and
Qammer Hussain Abbasi [6]

[1]   School of International Education, Xidian University, Xi'an 710071, Shaanxi, China; azizshahics2@gmail.com
[2]   School of Electronic Engineering, Xidian University, Xi'an 710071, Shaanxi, China;
      afren@mail.xidian.edu.cn (A.R.); dfan9308@gmail.com (D.F.); zhiyazhang@163.com (Z.Z.);
      nan_zhao_@hotmail.com (N.Z.); mluo@xidian.edu.cn (M.L.); fangming95@163.com (F.H.)
[3]   Northwest Women's and Children's Hospital, Xi'an 710061, Shaanxi, China; wwg530@163.com
[4]   School of Computer Science and Technology, University of Bedfordshire, Luton LU1 3JU, UK;
      masood.rehman@ieee.org
[5]   Electrical Engineering Department, University of Tabuk, Tabuk 71491, Saudi Arabia; obadarneh@ut.edu.sa
[6]   School of Engineering, University of Glasgow, Glasgow G12 8QQ, UK; Qammer.Abbasi@glasgow.ac.uk
*   Correspondence: xdyang@xidian.edu.cn; Tel.: +86-029-88202830

Received: 1 December 2017; Accepted: 19 March 2018; Published: 27 March 2018

Abstract: Medical healthcare is one of the fascinating applications using Internet of Things (IoTs).
The pervasive smart environment in IoTs has the potential to monitor various human activities by
deploying smart devices. In our pilot study, we look at narcolepsy, a disorder in which individuals
lose the ability to regulate their sleep-wake cycle. An imbalance in the brain chemical called orexin
makes the sleep pattern irregular. This sleep disorder in patients suffering from narcolepsy results in
them experience irrepressible sleep episodes while performing daily routine activities. This study
presents a novel method for detecting sleep attacks or sleepiness due to immune system attacks
and affecting daily activities measured using the S-band sensing technique. The S-Band sensing
technique is channel sensing based on frequency spectrum sensing using the orthogonal frequency
division multiplexing transmission at a 2 to 4 GHz frequency range leveraging amplitude and
calibrated phase information of different frequencies obtained using wireless devices such as card,
and omni-directional antenna. Each human behavior induces a unique channel information (CI)
signature contained in amplitude and phase information. By linearly transforming raw phase
measurements into calibrated phase information, we ascertain phase coherence. Classification and
validation of various human activities such as walking, sitting on a chair, push-ups, and narcolepsy
sleep episodes are done using support vector machine, K-nearest neighbor, and random forest
algorithms. The measurement and evaluation were carried out several times with classification
values of accuracy, precision, recall, specificity, Kappa, and F-measure of more than 90% that were
achieved when delineating sleep attacks.

Keywords: Internet of Things; S-band sensing; smart devices

---

1. Introduction

The Internet of Things (IoT) is a concept reflecting a connected set of anyone, anything, anytime,
anyplace, any service, and any network [1]. The Internet of Thing is a discussion and trend that has
been going on in the tech industry for a while. There are two big things to note about IoTs; first, the
average individual does not leverage it greatly. Second, IoTs takes analog ideas or devices, such as
sensors, actuators, home appliances etc., and connects them through inter-networking [2].

IoTs provides appropriate solutions for a wide range of applications such as smart cities, traffic congestion, waste management, structural health, security, emergency services, logistics, retails, industrial control, and health care [1].

Medical healthcare is one of the fascinating applications for IoTs. The pervasive smart environment leveraging IoTs has a potential to monitor various human's activities by the deployment of smart devices. Recently, in the medical healthcare domain, the research focus has moved toward human activity recognition. Daily routine activities have a prospect to be monitored and supported by IoTs. Patient's activity detection is a vital issue for caregivers with adequate information about the subject can be documented in healthcare systems. Activity recognition enables nurses or caretakers to provide timely assistance. For instance, in a pervasive environment, a caregiver can monitor activity using devices such as a computer or mobile phone to provide instant care for subject in risky situations [3]. For these reasons, we seek to incorporate small wireless devices used in S-band sensing technique to monitor the sleep attacks of patients suffering from narcolepsy disease (ND).

Narcolepsy is a neurological disorder in which individuals lose the ability to regulate their sleep–wake cycle. The normal cycle between sleeping and being awake is blurred, leading to sleeping excessively during the day [4]. It is an impairment with symptoms such as acute drowsiness and falling asleep without any prior warning. Loss of orexin neurons in the brain is attributed to the disruption of transitions between sleep–wake states [5]. People suffering from narcolepsy disorder have fewer excitatory neurons, and each neuron carries less of the neuropeptides orexin A and orexin B [6]. These orexins increase the activity of wake-promoting regions of the brain, thereby tipping the scales in favor of wakefulness and preventing inappropriate transitions into a sleeping state. ND primarily damages the neurons delivering orexin, and therefore less orexin is sent out and sleep-related symptoms start to intrude into wakefulness. ND can be triggered by activities such as walking, exercise, etc. or sudden, strong emotions like anger or excitement [7]. Narcolepsy sleep episodes typically last up to few minutes.

In a nutshell, patients suffering from narcolepsy fall asleep involuntary while doing daily activities. Continuous monitoring of ND patients, for targeted treatment—particularly the elderly—is of the utmost importance to avoid the risk of injury or to shed light on its triggers [5]. Thus, to monitor the daily routine activities of ND patients, we use the S-Band sensing technique leveraging wireless devices such as card, omni-directional antenna, etc. Each human activity generates a particular wireless channel information imprint in terms of variances of amplitude and phase information that is used to detect various human activities for sleep attack experiences by ND patients. The amplitude and phase information obtained are classified using support vector machine (SVM), K-nearest neighbor (KNN), and random forest (RF) for identifying sleep attacks.

2. Related Work

Human activity recognition for various applications has attracted much attention from researchers around the world. Applications include fall detection [8], activity detection for energy saving at homes or offices [9], 24-hour sleep-wake monitoring in narcolepsy [10], a detection system for motion disorders in Autism patients [11], and other uses leveraging IoTs [12–18]. The methods introduced in [13,14] leverage body sensor nodes powered by human energy harvesting and wireless sensor networks for remote patient monitoring. Cretikos et al. [19], Pantelopoulos et al. [20], and Coronato et al. [14] introduced systems for monitoring vital signs in medical problems. Many chronic diseases as in [20,21] can be forecasted and prevented using activity recognition. However, recording such information in a timely manner is a daunting task. Several methods have been proposed both in academia and in the industry for healthcare informatics and can be categorized as non-invasive sensors [22–24], and invasive sensors [25–27]. Currently, invasive sensors are a rich source of information as far as healthcare is concerned. However, some invasive sensors are uncomfortable and not suitable for everyone since patients suffering from Parkinson's disease, narcolepsy, skin diseases, and infants are not encouraged for wearing such sensors [22] or to be deployed on their bodies. Qi et al. [27] propose

RadioSense, a prototype system of ZigBee radio-based activity sensing leveraging Institute of Electrical and Electronics Engineers (IEEE) 802.15.4 specifications. Thus, the non-invasive sensors might be the only solution for patients with the aforementioned diseases. Various technologies have successfully demonstrated the efficacy of non-invasive sensors. Kaushik et al. [28] proposed a non-invasive system leveraging a pyroelectric infra-red detector. Zhou et al. [29] developed video-based system detecting various activities with good classification accuracy but raised privacy concerns. Tsutsui [23] utilized ultrasound echo to detect different human movements, but the system provided poor coverage range. Li et al. [30] used Doppler radar for fall detection and gait analysis with adequate performance. Zhang et al. [31] exploited an ultra-wide-band (UWB) radar system for vital signs monitoring. However, the aforementioned technique leveraging radar faces limitations such as infrastructure deployment, spectrum licensing, and the fact that it requires specialized hardware. This paper presents a novel method leveraging the S-Band sensing technique using IEEE 802.11 specifications operating at 2.4 GHz exploiting wireless devices such as a card, antennas, etc.

3. S-Band Sensing and Data Processing

The proposed method uses the S-Band sensing technique for tracking human activities. The S-Band sensing technique exploits wireless channel information to obtain signals by use of wireless devices such as an omni-directional antenna as a receiver, and a card that is connected to the receiving antenna.

The S-Band sensing technique is channel sensing based on frequency spectrum sensing using the orthogonal frequency division multiplexing transmission at 2 to 4 GHz frequency range. The transceiver has higher efficiency because it uses frequencies that are orthogonal increasing the robustness. The orthogonal frequency division multiplexing is composed of 64 frequencies spaced 312.5 KHz apart [32]. The spacing is selected because of the FFT sampling size. The IEEE 802.11n uses 52 frequencies for data, eight as null frequencies, and four as pilot carriers over a channel. The card primarily reports 30 different frequencies for data processing and form one channel information (CI) packet. The main advantage of obtaining multiple frequencies is that any one or more than one frequencies can be used for analysis.

The wireless channel parameters such as carrier frequency, bandwidth, delay spread, and Doppler spread are extremely sensitive to variations in the spatial domain, causing multipath fading. We thus use such CI data that are affected by multipath reflections generated by human motion in a room or free space with a transmitting–receiving antenna. To discern the signals amplitude and phase information unique to motion, we need a second level of processing. The variances of amplitude and phase information for activity recognition generated by motion have to be looked at for uniqueness in their CI signatures.

The method we present efficiently detects various human activities such as walking, squatting, sitting on chair, pushups, and narcolepsy sleep attacks using the S-band sensing technique, which can be reported in a human understandable format. The raw phase information retrieved via the network interface card is extremely random and inapplicable for the detection of sleep episodes. Hence, the proposed design fully exploits both amplitude and calibrated phase information by applying a linear transformation on the raw phase data discussed in the subsequent section. Furthermore, SVM, KNN, and RF classifiers are used to segregate human activities and detect sleep episodes. The performance metrics such as accuracy, precision, recall, specificity, Kappa, and F-measure were used to evaluate the three classifiers.

An RF signal is received from a transmitter through multiple paths in an indoor environment when the transceiver pair is connected via network interface card. The received signal can be expressed as

$$Y_j = X_j \times WCI_j + N, \tag{1}$$

Here, Y and X denote the received and transmitted signals, respectively, for jth frequency. The CI data is the wireless channel information; that carries the channel frequency response (CFR) and N is the random noise. The raw CI data can be denoted as follows:

$$WCI_j = \left\|WCI_j\right\| e^{j\angle WCI_j},\tag{2}$$

where $\|WCI_j\|$ represents the amplitude information and $\angle WCI_j$ is the phase data for all $j = [1, 30]$. A packet or channel matrix specifies the amplitude and phase information measured for a certain period of time. The wireless channel information over a time period for a particular frequency can be expressed as

$$WCI_{time\_history} = WCI_j^1 + WCI_j^2 + WCI_j^3 \dots WCI_j^z,\tag{3}$$

Here, j is the particular frequency number and z is the total number of CI packets received.

Phase Calibration

The phase data extracted from wireless CI stream is largely random and inapplicable for detecting Narcolepsy sleep episodes due to the presence of random noise and phase offset between transceiver pair. Thus, to mitigate the impact of random noise from the CI phase data and remove the phase offset, we use linear transformation on raw phase data.

Let $\angle WCI^\wedge_j$ represent the measured wireless CI phase data for jth frequency.

$$\angle WCI^\wedge_j = \angle WCI_j + 2\pi \frac{k_j}{Z}\tau + \beta + N,\tag{4}$$

Here, $\angle WCI$ is the true phase data, N is the random noise, τ is the unsynchronized timing between transceiver, and β is the unknown phase offset. The frequency indices k_j for $j = [1, 30]$ and FFT size Z can be obtained from IEEE 802.11n specifications [33]. The unknown terms τ and β listed above make it infeasible to obtain useful phase information solely from S-band sensing technique. Thus, to remove the unknown terms, we apply a linear transformation to the raw CI phase data across the frequency bands [34].

Let the two terms L and M represent the slope of the CI phase and phase offset considering overall frequency bandwidth, respectively.

$$L = \frac{\angle WCI^\wedge_{30} - \angle WCI^\wedge_1}{k_{30} - k_1}.\tag{5}$$

$$M = \frac{1}{30}\sum_{i=1}^{30} \angle WCI^\wedge_j.\tag{6}$$

By subtracting the linear term $Lk_j + M$ from raw CI phase data, we get the calibrated phase information $\angle WCI^\circ_j$ [10], denoted as

$$\angle WCI^\circ_j = \angle WCI^\wedge_j - Lk_j - M.\tag{7}$$

4. Data Classification

Data classification is the process of categorizing data sets into different forms, types or any other distinct class. There exist several data classification techniques such as K-nearest neighbor (KNN) [35], neural networks (NN) [36], Random Forest (RF) [37], and support vector machine. In this paper, we have compared several data classification techniques such as the support vector machine, KNN, and RF algorithm.

4.1. Support Vector Machine for Data Classification

The advantage of using the SVM algorithm is that it provides efficient performance when it comes to practical problems [38,39]. Also, the inner-product kernel functions of SVM solve the linearly non-separable problems in higher dimension space.

The support vector machine is applied to classify subject's various activities such as walking, squatting, push-ups, sitting on a chair, and sleep episodes due to ND. The SVM is applied to the measured CFR data received using the S-band sense. SVM is a binary classifier developed for non-linear boundary problems [40]. A hyper-plane is used as a decision boundary to classify the two datasets. The closest data points to the hyper-plane, which impart construction of the hyper-plane, are called support vectors [41]. The optimum hyper-plane can be written as follows:

$$w^T x + c = 0, \tag{8}$$

here w denotes the weight vector, x is the input vector and c is the bias. Support vectors representing each class are expressed as

$$w^T x + c = 1, \text{ for }; \text{ corresponds to class A}$$

$$w^T x + c = -1, \text{ for } y_i = -1; \text{ corresponds to class B}$$

The optimum hyper-plane for training sample data is expressed as

$$\min \phi(w) = \frac{w^T w}{2}, \text{ for } y_i(w^T x_i + c) \geq 1 \tag{9}$$

where $i = 1, 2, 3, \ldots, n$. The inner product kernel functions used in this paper are shown in Table 1.

Table 1. Inner product kernel functions.

Type	Kernel Function $K(x, x_i), i = 1, 2, 3, \ldots, P$
Linear	$x^T x_i + c$
Polynomial	$\left(x^T x_i + 1\right)^p$
Radial-basis function (RBF)	$e^{\left(\frac{-\|x - \|x_i\|\|^2}{2\sigma^2}\right)}$

Since we need to differentiate the ND sleep episodes from other four human activities, we adopt a one-versus-rest approach using SVM. The CFR data comprising amplitude information for the individual frequency and calibrated phase information is discussed in Section 3. Data obtained using S-band sensing techniques were processed to extract features for classifying various human activities accurately. The feature selection technique is important to reduce the number of operations and to represent the key changes against the amplitude or phase.

Various human activities can be classified using time-domain analysis, frequency-domain analysis, and time-frequency domain analysis [39]. The time-domain features can be easily extracted since they are the simplest ones, as compared to the frequency-domain and time-frequency domain features. In time-domain approaches, the time-domain waveform is analyzed directly to extract statistical indices such as the root-mean-square amplitude, skewness, and kurtosis [31]. Frequency-domain approaches are usually employed to find the characteristic frequencies via frequency analysis, such as the Fourier spectrum, cepstrum analysis, and the envelope spectrum [42]. This approach is also simple and characterizes an intuitive nature by allotting the components to their corresponding frequency in a particular spectrum. Time-frequency domain approaches including wavelet analysis, the fast Fourier transform (FFT), Wigner–Ville distribution, and Hilbert–Huang transform, etc., investigate waveform signals in both the time and frequency domain and can provide more information about the data classification [43,44]. However, this approach is essentially more complicated than the frequency-domain or time-domain approaches in a practical scenario. In this study, we use 10 time-domain statistical features as listed in Table 2. Here, x_i denotes a particular frequency of choice while x denotes the signal of every frequency.

Table 2. Ten features used to train and test support vector machine (SVM).

Root mean square $(Y_{RMS}) = \sqrt{\frac{1}{P}\sum_{i=1}^{P} x_i^2}$	Crest Factor $(Y_{CF}) = \frac{\max(|x_i|)}{Y_{rms}}$
Marginal factor $(Y_{MF}) = \frac{\max(|x_i|)}{Y_{SRA}}$	Skewness value $(Y_{SV}) = \frac{1}{P}\sum_{i=1}^{P}\left(\frac{[x_i-\mu_x]}{\sigma}\right)^3$
Square root of amplitude $(Y_{SRA}) = \left[\frac{1}{P}\sum_{i=1}^{P}\sqrt{|x_i|}\right]^2$	Impact factor $(Y_{IF}) = \frac{\max(|x_i|)}{\frac{1}{P}\sum_{i=1}^{P}|x_i|}$
Mean value $(Y_{MV}) = \frac{1}{N}\sum_{i=1}^{P} x_i$	Peak to peak value $(Y_{PPV}) = \max(x_i) - \min(x_i)$
Kurtosis value $(Y_{KV}) = \frac{1}{P}\sum_{i=1}^{P}\left(\frac{[x_i-\mu_x]}{\sigma}\right)^4$	Standard deviation $(Y_{STD}) = \sqrt{\frac{1}{P}\sum_{i=1}^{P}(x_i-\mu_x)^2}$

4.2. K-Nearest Neighbor Algorithm

K-Nearest Neighbor (KNN) is a simple data classification method that takes k number of nearest training samples in the feature space. KNN algorithm works on a dataset that is classified by its neighbors based on majority vote. The dataset is assigned to a particular class that is common in the k nearest neighbor defined by the distance function. For example, when $k = 1$, the dataset is assigned to the nearest neighbor class. There are three main distance functions used in KNN algorithm considering continuous variables.

$$Manhattan = \sum_{i=3}^{k}|x_i - y_i| \tag{10}$$

$$Minkowski = \left(\sum_{i=1}^{k}|x_i - y_i|^q\right)^{1/q} \tag{11}$$

$$Euclidean = \sqrt{\sum_{i=1}^{k}(x_i - y_i)^2} \tag{12}$$

$$Hamming\ Distance = D_H = \sum_{i=1}^{k}|xi - yi| \tag{13}$$

In a case where categorical variables exist, Hamming distance as in Equation (13) should be used. When there is a combination of numerical values and categorical variables in a given dataset, the issue of standardization arises as the numerical variables are between 0 and 1. The optimum value for k is chosen by thorough inspection of the given datasets. Generally, when the value of k is large, the overall noise is reduced. Cross validation or rotation estimation is another method to identify the optimal value for k.

4.3. Random Forest Algorithm

Random forest algorithm is another power machine learning algorithm that is capable of performing classification tasks. RF algorithm is combination of different but simple predictors that can decrease the computational cost and obtain better performance [45]. Employing the regression tree as the learning algorithm of sub-models, the random forest is a modified version of Bagging [45]. In random forest algorithm, the random selection features grow a tree on the new training datasets at every node to identify the split. The RF algorithm reduces the complexity in following ways:

the parallel computing procedure in combination with parallel ensemble, and the introduction of the sub-model at each sample subset with no communication from the CPU [45]. The construction of the sub-models built on the subsets dramatically decreases the training samples. The bootstrap replications, such as sample subsets, are simple and maintain very low complexity. Also, the learning algorithm of the sub-models is considered to be the regression tree.

5. Experimental Setup

We implemented the experiment in a large room (25 m × 25 m) as shown in Figure 1. The experiment had two parts. In the first part, microwave-absorbing materials were deployed along the walls. Three tables, three chairs, and two persons at one time, one taking the measurements and a subject, were present inside the room, as shown in Figure 1a. The main reason for using microwave-absorbing material was to reduce multipath reflections that could affect the experimental results. In the second part, as indicated in Figure 1b, the experimental design was the same, but the microwave-absorbing material was removed, and it was observed that the percentage accuracy (described in Table 3) slightly decreased due to multipath propagation. S band antenna was deployed as a transmitter, and an omni-directional antenna served as a receiving antenna. The transmitter and receiver were placed 15 m apart at a 1-m height. The receiving antenna worked as the detection point. A channel information collection tool introduced by Harpin [33] was installed in the host computer would constantly receive 20 wireless CI packets per second. In order to differentiate the ND episodes from other daily activities, we tested five body motions: walking, squatting, push-ups, sitting on chair, and subject falling asleep due to narcolepsy. Each body motion disturbed the wireless medium; as a result, a unique CI signature was recorded, indicating a particular human activity. The experiments were performed for 16 ND patients.

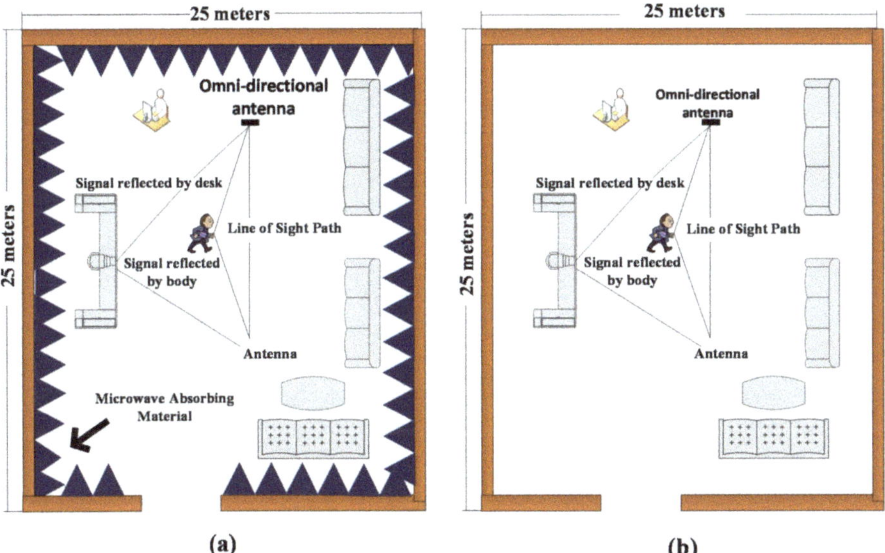

Figure 1. Experiment design for detection sleep attacks. (**a**) Experimental setup with the microwave absorbing material. (**b**) Experimental setup without microwave absorbing material.

Table 3. Accuracy of SVM used for sleep attack detection (%).

(a) Accuracy of SVM Used for Sleep Attack Detection—With Microwave Absorbing Material											
Kernel		**5 Features**					**10 Features**				
Function	**S**	**a**	**b**	**c**	**d**	**e**	**a**	**b**	**c**	**d**	**e**
Linear	40	98.75	78.25	90.50	98.25	65.00	97.00	75.25	94.75	98.50	84.00
	80	98.25	77.25	91.75	98.00	65.50	98.75	78.25	95.25	98.75	87.00
	120	98.25	76.50	90.00	98.00	66.25	99.00	78.00	95.50	99.00	89.75
Polynomial	40	98.50	69.25	89.50	95.00	80.50	95.50	81.75	90.50	98.50	89.75
	80	99.50	71.50	87.75	95.00	85.25	98.00	87.75	92.25	98.50	92.00
	120	99.00	74.25	88.50	99.00	87.75	98.50	84.75	95.25	98.50	92.75
RBF	40	98.25	74.75	89.50	98.00	84.00	98.25	78.75	92.50	99.50	88.50
	80	98.25	74.50	89.00	98.25	86.25	98.25	83.25	95.00	99.25	91.00
	120	98.25	88.00	88.50	98.00	87.25	98.75	84.50	96.00	99.25	91.25

(b) Accuracy of SVM Used for Sleep Attack Detection—Without Microwave Absorbing Material											
Kernel		**5 Features**					**10 Features**				
Function	**S**	**a**	**b**	**c**	**d**	**e**	**a**	**b**	**c**	**d**	**e**
Linear	40	91.65	74.19	85.50	96.80	60.00	91.50	73.13	91.98	96.52	82.22
	80	93.11	74.87	86.55	96.14	61.43	92.04	74.91	92.11	96.61	84.74
	120	95.55	74.71	87.52	96.00	62.31	91.95	75.11	97.50	97.43	88.75
Polynomial	40	92.75	66.14	88.72	96.61	77.52	92.50	79.81	88.00	96.19	81.12
	80	94.28	66.50	89.85	91.21	79.52	94.42	79.90	89.13	95.93	89.54
	120	94.96	70.13	88.89	97.11	83.81	96.62	80.01	91.72	97.09	88.75
RBF	40	96.69	71.11	90.10	98.21	80.14	97.15	76.75	90.80	95.31	85.77
	80	96.11	71.32	90.32	98.71	80.85	97.95	80.94	91.84	95.54	90.64
	120	96.25	74.43	90.51	98.28	81.25	97.75	81.74	92.56	96.33	88.91

6. Results and Discussion

This section describes the simulation results obtained during the experiment. Firstly, we examine the CFR obtained for five human activities such as sitting on a chair, walking around the room, push-ups, squatting, and sleep episodes due to ND, as shown in Figure 2.

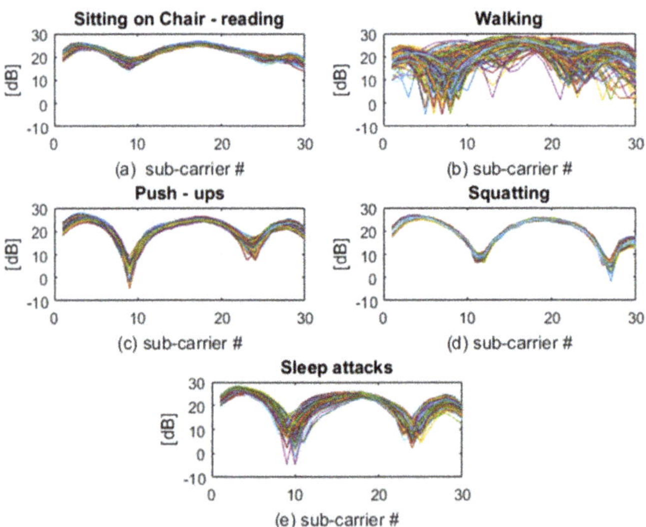

Figure 2. Channel frequency response (CFR) data obtained for five different human activities, different colors in the figures indicate variances of amplitude information of different frequencies.

Figure 2 describes the CFR data for various human activities, transitioning from one activity to another over a period of time. We analyze each human activity that has induced a particular CFR pattern. For example, when the subject was sitting on a chair, the corresponding amplitude information against 30 frequencies is almost stationary, as evident in Figure 2a, but as the person started walking, the wireless medium constantly got disturbed, and as a result, the CFR data also changed, as in Figure 2b. On the other hand, slight variances in amplitude information against the 30 frequencies can be observed in Figure 2c when the subject was performing push-ups on the ground. Figure 2e describes the results of squatting. When the person experienced sleep episodes, due to which no large-scale movement was observed within wireless range, the amplitude information remained stationary. To further validate our point, we looked into the CI signatures generated by various activities considering the calibrated phase information and analyzed the data shown in Figure 3.

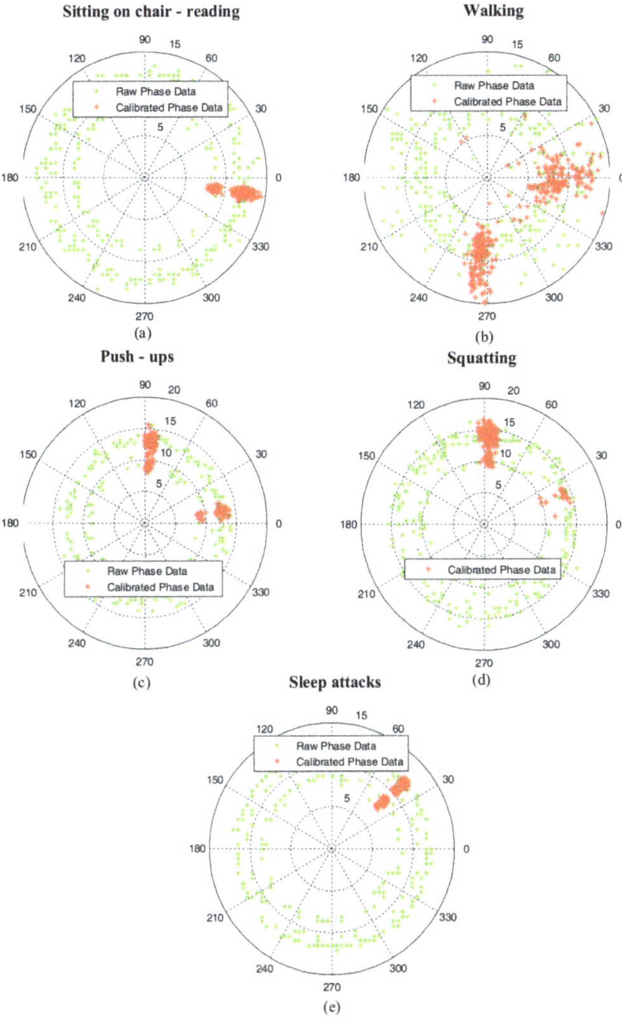

Figure 3. Raw channel information (CI) phase and calibrated phase information obtained during the experiment.

Analyzing the calibrated phase information for corresponding human activities, we can clearly see that each human behavior has generated a particular registration as far as calibrated phase is concerned. A cluster of stationary data can be seen in Figure 3a that indicate sitting. Looking at activities such as motion or sitting, we are able to differentiate them with raw data and also with an SVM. Unlike a sitting posture, significant variances in the calibrated CI phase information can be seen in Figure 3b when a person starts walking within the area of interest. When the subject makes the transition to push-ups, two chunks of data can be seen in Figure 3c. Similarly, the phase information changes when the person is squatting. Finally, Figure 3e shows the stationary data when a person experienced sleepiness. In order to discuss the time history of narcolepsy sleep attacks, we analyze the acquired signal, as seen in Figure 4.

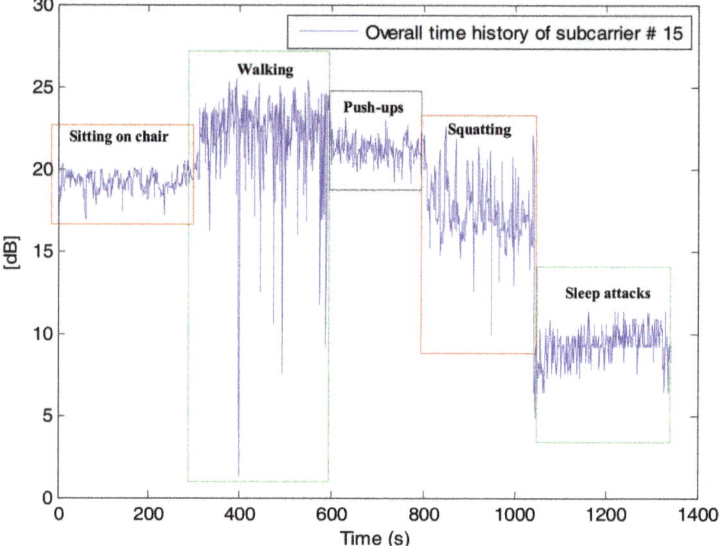

Figure 4. Time history of person's five different activities.

Figure 4 shows that 1400 packets in total were received for 1400 s when the experiment was being performed. A frequency # 15 was selected for amplitude-vs.-time history analysis as in [44]. We can see that the amplitude levels for sleep episodes fluctuate between 5 dB to 10 dB from 1150 s to 1350 s are clearly different from the other four activities. However, the power levels for walking and push-ups remain the same for a certain period of time. Similarly, there are moments where the power level of push-ups, walking and sitting activity are similar. Thus, to classify the different human activities and detect the ND sleep episodes accurately, we use the support vector machine.

6.1. Classification Results

6.1.1. Results Obtained Using SVM

As discusses in an earlier section, the proposed method utilizes two parameters for evaluation: the variances of amplitude information and calibrated phase information when the subject performs various activities. We adopted the one-versus-rest approach and trained the SVM classifier using built-in function in MATLAB called "svmtrain" and obtained the optimal decision boundary by utilizing the sequential minimal optimization solver. In order to train the SVM, 40, 80, and 120 samples from each class were used separately. The 10 SVM features extracted from training data are indicated in Table 2. Each one of them describes the measurement per CI packet distinctively. The results in this particular table indicate the percentage accuracy, which is one of the six performance metrics used in this paper. We have explored the remaining metrics along with accuracy in a later part of this paper. The data classification results using the SVM algorithm are described in Table 3 using microwave absorbing material and without using microwave absorbing material, respectively. The first column in Table 3 indicates the type of SVM kernel function used for training data set, and the second column denotes the training samples (40, 80 and 120). The remaining five columns *a, b, c, d, e* representing the sitting, walking, push-ups, squatting, and sleep attacks describes the percentage accuracy. The results indicate that when the linear kernel function was used to differentiate the five human activities from 40 training samples, the percentage accuracy is 98.75% for sitting, 78.25% for walking, 98.25 for push-ups, and 65% for sleep attacks. Similarly, the accuracy rate is around 65% for sleep attacks using linear kernel function when 80 and 120 training samples were considered, as shown in Table 3. An increase in the training dataset marginally improved the accuracy rate for sleep attacks but remained around 65%. However, the percentage accuracy greatly improved when the ten SVM features were considered using linear kernel function. The accuracy is between 75.25% and 98.5% for 40 training samples, 78.2% to 98.75% for 80 training samples, and 78% to 99% for 120 training samples. The accuracy rate of sleep attacks detection for 40, 80, and 120 training samples is 84%, 87%, and 89.75%, respectively, as shown in Table 3. The polynomial kernel function worked better than linear kernel function. The error rate is between 69.25% and 98.5% for 40 training samples, 71.5% to 99.5% for 80 training samples, and 74.25% to 99% for 120 training samples when the polynomial kernel function was used as far as five SVM features were considered. The accuracy rate obtained for the RBF kernel function is between 74% and 98% when five SVM featured were used, as shown in Table 3. However, the polynomial kernel function and RBF kernel function provided the best results when 10 SVM features were used. The accuracy rate obtained using polynomial kernel function considering ten SVM features is between 81% to 98.5% for 40, 80, and 120 training samples, respectively. While the percentage accuracy is between 78.75% and 98.75% when 10 SVM features were used, considering the RBF kernel function. However, we observe a slight decrease in percentage accuracy when the experiment was performed without microwave-absorbing material. The results obtained are shown in Table 3. Table 3 indicates the comparison between the three SVM kernel functions used for transforming the data. We observe that the accuracy rate is decreased by 5–7% without deploying microwave-absorbing material in the indoor environment. We further investigate the results obtained using KNN and RF algorithms.

6.1.2. Data Classification Using KNN and RF

The data collected using S-Band sensing technique was then classified using KNN and RF algorithms [45–47]. The former classification method is based on supervised learning where voting criteria classifies the new objects. The nearest object k from the training datasets is considered that assigns a new object to a particular class based on the majority votes. The training procedure of the KNN classification method includes storing of features and class label of the particular training datasets. In classification procedure, the most frequent training samples k assign an unlabeled object of a particular class. Various distance parameters are used in KNN algorithm as discussed in Equations (10)–(13). In our case, we have used Euclidean distance as in Equation (12) and set the value of k to 1, which implies that the selected class label was exactly the same as the one nearest to the training dataset On the other hand, the RF classification method primarily integrates a set of independent decision tree classifiers [48]. A decision tree in an RF classifier with N number of leaves divides the feature space into N number of regions such as $R_m, 1 \leq m \leq N$. The prediction function $f(x)$ for each tree is described as

$$f(x) = \sum_{m=1}^{N} c_m \Im(x, R_m),$$

Here, N indicates the total number of regions in a feature space, R_m corresponds to the m region, c_m represents a constant number that corresponds to m, and $\Im$ describes the indicator function.

$$\Im\,(x, R_m) = \left(\begin{array}{l} 1,\ if\ x\ \varepsilon R_m \\ 0,\ otherwise \end{array}\right.$$

The final decision function for random forest algorithm is based on majority vote of all the trees. In our case, we have used an RF classification method with 100 trees.

We now discuss the data obtained using KNN and RF algorithms in terms of accuracy, precision, recall, specificity, Cohen's Kappa coefficient, and F-measure. The formulae for six performance metrics are discussed as follows:

$$Accuracy = \frac{Tp + Tn}{Tp + Tn + Fp + Fn} \tag{14}$$

Accuracy is a fraction of instances that are correctly classified [46],

$$Precision = \frac{Tp}{Tp + Fp} \tag{15}$$

Precision is estimated as a ratio of true predictions pertaining to a class [46].

$$Recall = \frac{Tp}{Tp + Fn} \tag{16}$$

Recall is the fraction of relevant instances that have been retrieved over the total amount of relevant instances [47].

$$Kappa = \frac{(Accuracy - Accuracy_{expected})}{1 - Accuracy_{expected}} \tag{17}$$

Kappa is the accuracy of the system to the accuracy of the random system.

$$Specificity = \frac{Tn}{Fp + Tn}, \tag{18}$$

Specificity measures the proportion of negatives that are correctly identified.

$$F - measure = \frac{2Tp}{(2Tp + Fp + Fn)} \tag{19}$$

F-measure is the harmonic mean of precision and recall.

6.1.3. Results Obtained Using SVM, KNN, and RF Classifiers

The KNN and RF classifiers were trained and tested using 120 samples. The confusion matrices obtained for SVM, KNN, and RF classifiers are shown in Table 4, respectively.

Based on these confusion matrices, we have obtained the values of six performance metrics as shown in Table 5.

Table 5 shows the accuracy, precision, recall, specificity, F-measure, and Kappa values obtained using support vector machine. In the earlier section, we discussed the percentage accuracy using three kernel functions considering 10 SVM features. The accuracy presented in Table 5 is obtained using the RBF kernel function considering 10 SVM features, as shown in Table 3. The precision for all five human activities is more than 90%, while for sleep attack, it is 94%. The recall values received using SVM vary between 91% and 94.4%. The specificity obtained is more than 96% for all activities. The F-measure values also vary between 90% and 95.4%. The average Cohen's Kappa coefficient value obtained for all five human activities is 0.925. Table 5 indicates the six performance metrics received using KNN classifier. The accuracy for sitting, walking and squatting is 94.8%, 92.8%, and 90.2%, respectively.

The accuracy rate for push-ups, and sleep attack is 86.6% and 88.5%, respectively. The precision values for walking and squatting are below 90%, while the rest of the three are above 90%. The recall values for push-ups and sleep attacks are 86.6% and 88.6%, respectively. The other three values are above 90%. The specificity is more than 90% for all five human activities. The F-measure values for all human activities are near to 90% except squatting activity (87%). The average Cohen's Kappa coefficient for all five activities is 0.811.

Table 4. Classification results considering different approaches.

(a) Confusion matrix obtained using SVM for 120 training samples					
	Sitting	Walking	Push-ups	Squatting	Sleep attacks
Sitting	118	3	1	1	2
Walking	1	101	1	0	1
Push-ups	0	4	116	0	3
Squatting	1	5	2	119	4
Sleep attacks	0	7	0	0	110
(b) Confusion matrix obtained using KNN for 120 training samples					
Sitting	111	1	1	3	1
Walking	2	104	2	4	0
Push-ups	4	6	110	6	1
Squatting	2	5	2	102	2
Sleep attacks	1	4	5	5	116
(c) Confusion matrix obtained using RF for 120 training samples					
Sitting	105	8	2	3	2
Walking	5	96	3	5	1
Push-ups	4	8	112	1	2
Squatting	3	3	2	109	2
Sleep attacks	3	5	1	2	113

Table 5. Classification results..

(a) Classification results obtained using SVM (%)					
	Accuracy	Precision	Recall	Specificity	F-Measure	Kappa
Sitting	98.3	94.4	94.4	99.0	96.0	
Walking	84.1	97.1	97.0	96.0	90.0	
Push-ups	96.6	94.3	94.3	99.1	95.4	0.925
Squatting	99.1	90.8	91.0	99.7	95.2	
Sleep attacks	91.6	94.0	94.0	97.8	92.8	
(b) Classification results obtained using SVM (%)						
Sitting	94.8	92.5	94.8	97.5	93.6	
Walking	92.8	86.6	92.8	96.6	89.6	
Push-ups	86.6	91.6	86.6	97.7	89.0	0.811
Squatting	90.2	85.0	90.2	96.0	87.5	
Sleep attacks	88.5	96.6	88.5	99.0	92.4	
(c) Classification results obtained using RF (%)						
Sitting	87.5	87.5	87.5	96.6	84.4	
Walking	87.2	80.0	87.2	94.8	83.4	
Push-ups	88.1	93.3	88.1	97.5	90.6	0.865
Squatting	91.5	90.8	91.5	97.4	91.2	
Sleep attacks	91.1	94.1	91.1	98.3	92.6	

We further examine the results obtained using random forest algorithm as shown in Table 5. The accuracy rate for sitting and walking is approximately 87%. For push-ups, the accuracy is 88.1%, while for squatting and sleep attacks, the accuracy rate is approximately 91%. Looking at the precision values, sitting has a received value of 87.5%, walking has 80.0%, and the rest have received values of more than 90%. The recall value for sitting and walking is approximately 87%, the push-ups value is 88.1%, and for squatting and sleep attacks, the value is approximately 91%. The specificity for all five human activities is more than 94%. The F-measure value for sitting and walking activity is 84.4% and 83.4%, respectively, while the value for value is more than 90% for the rest of the three activities. The average Cohen's Kappa coefficient for all five human activities is 0.865%.

When we compare the values of all the six performance parameters, it is observed that the SVM algorithm provides the best results as compared to the other two classification techniques. All the values of the six performance metrics considering five human activities are more than 90%, except for of walking, which is 84.1%. As far as the KNN and RF classifiers are concerned, there are multiple instances where the values of the performance metrics are below 90%.

7. Conclusions

This paper presented an automatic method for healthcare applications in IoTs that continuously monitors a patient's different activities and reports sleep episodes due to narcolepsy. The method leveraged S-band sensing technique is used to record variances of wireless channel information. We examined the variances of amplitude and calibrated phase information of CI data and observed that each human activity induced a particular CI registration. Three classification algorithms such as SVM, KNN, and RF algorithms were used to classify different human activities and identify sleep episodes. It was observed that the SVM classifier provided better results than KNN and RF classifiers by examining the six performance metrics. The proposed method is easily deployable and cost-effective. A musical beat alarm helps the patient in waking up and restoring routine activities.

Acknowledgments: The work was supported in part by the Fundamental Research Funds for the Central Universities (JB180205 and JB160211), in part by the National Natural Science Foundation of China (Grant No. 61671349, 61301175, 61601338), in part by the International Scientific and Technological Cooperation and Exchange Projects in Shaanxi Province (Grant No. 2017KW-005). The authors would like to thank Northwest Women's and Children's Hospital for its support.

Author Contributions: Syed Aziz Shah, Aifeng Ren, Dou Fan, Zhiya Zhang, Nan Zhao, Xiaodong Yang, Ming Luo designed the experiments; Syed Aziz Shah, Dou Fan performed the experiments; Syed Aziz Shah, Dou Fan analyzed the data; Weigang Wang, Fangming Hu, Masood Ur Rehman, Osamah S. Badarneh, Qammer Hussain Abbasi contributed the materials; Syed Aziz Shah wrote the paper.

Conflicts of Interest: The authors declare no conflict of interest.

References

1. Islam, S.R.; Kwak, D.; Kabir, M.H.; Hossain, M.; Kwak, K.S. The internet of things for health care: A comprehensive survey. *IEEE Access* **2015**, *3*, 678–708. [CrossRef]
2. Kumar, N.; Rodrigues, J.J.P.C.; Chilamkurti, N. Bayesian coalition game as-a-service for content distribution in internet of vehicles. *IEEE Internet Things J.* **2014**, *1*, 544–555. [CrossRef]
3. Tentori, M.; Favela, J. Activity-aware computing for healthcare. *IEEE Pervasive Comput.* **2008**, *7*. [CrossRef]
4. Kudo, M.; Sklansky, J. Comparison of algorithms that select features for pattern classifiers. *Pattern Recognit.* **2000**, *33*, 25–41. [CrossRef]
5. Burgess, C.R.; Scammell, T.E. Narcolepsy: neural mechanisms of sleepiness and cataplexy. *J. Neurosci.* **2012**, *32*, 12305–12311. [CrossRef] [PubMed]
6. La Herrán-Arita, D.; Alberto, K.; Guerra-Crespo, M.; Drucker-Colin, R. Narcolepsy and Orexins: An example of progress in sleep research. *Front. Neurol.* **2011**, *2*, 26. [CrossRef] [PubMed]
7. Siddiqui, M.M.; Srivastava, G.; Saeed, S.H. Diagnosis of narcolepsy sleep disorder for different stages of sleep using Short Time Frequency analysis of PSD approach applied on EEG signal. In Proceedings of the Computational Techniques in Information and Communication Technologies (ICCTICT), New Delhi, India, 11–13 March 2016; pp. 500–508.
8. Han, C.; Wu, K.; Wang, Y.; Ni, L.-M. Wifall: Device-free fall detection by wireless networks. *IEEE Trans. Mob. Comput.* **2017**, *16*, 581–594.

9. Pu, Q.; Gupta, S.; Gollakota, S.; Patel, S. Whole-home gesture recognition using wireless signals. In Proceedings of the 19th Annual International Conference on Mobile Computing & Networking, Miami, FL, USA, 30 September–4 October 2013; pp. 27–38.

10. Kohsaka, M.; Fukuda, N. Twenty-four-hour sleep-wake monitoring in narcolepsy: Comparison with MSLT. *Sleep Med.* **2013**, *14* (Suppl. 1), e172. Available online: https://doi.org/10.1016/j.sleep.2013.11.403 (accessed on 5 March 2018). [CrossRef]

11. Coronato, A.; de Pietro, G.; Paragliola, G. A situation-aware system for the detection of motion disorders of patients with Autism Spectrum Disorders. *Expert Syst. Appl.* **2014**, *41*, 7868–7877. [CrossRef]

12. Islam, M.Z.; Nahiyan, K.M.T.; Kiber, M.A. A motion detection algorithm for video-polysomnography to diagnose sleep disorder. In Proceedings of the 2015 18th International Conference on Computer and Information Technology (ICCIT), Dhaka, Bangladesh, 21–23 December 2015; pp. 272–275.

13. Ibarra, E.; Antonopoulos, A.; Kartsakli, E.; Rodrigues, J.J.; Verikoukis, C. QoS-aware Energy Management in Body Sensor Nodes Powered by Human Energy Harvesting. *IEEE Sens.* **2016**, *16*, 542–549. [CrossRef]

14. Tennina, S.; Santos, M.; Mesodiakaki, A.; Mekikis, P.V.; Kartsakli, E.; Antonopoulos, A.; Di Renzo, M.; Stavridis, A.; Graziosi, F.; Alonso, L.; et al. WSN4QoL: WSNs for Remote Patient Monitoring in e-Health Applications. In Proceedings of the IEEE ICC, Kuala Lumpur, Malaysia, 22–27 May 2016.

15. Jara, A.J.; Zamora, M.A.; Skarmeta, A.F. An internet of things—Based personal device for diabetes therapy management in ambient assisted living (AAL). *Pers. Ubiquit. Comput.* **2011**, *15*, 431–440. [CrossRef]

16. Tian, J.; Morillo, C.; Azarian, M.H.; Pecht, M. Motor Bearing Fault Detection Using Spectral Kurtosis-Based Feature Extraction Coupled With K-Nearest Neighbor Distance Analysis. *IEEE Trans. Ind. Electron.* **2016**, *63*, 1793–1803. [CrossRef]

17. Dong, B.; Ren, A.; Shah, S.A.; Hu, F.; Zhao, N.; Yang, X.; Haider, D.; Zhang, Z.; Zhao, W.; Abbasi, Q.H. Monitoring of atopic dermatitis using leaky coaxial cable. *Healthc. Technol. Lett.* **2017**, *4*, 244–248. [CrossRef] [PubMed]

18. Yang, X.; Shah, S.A.; Ren, A.; Zhao, N.; Fan, D.; Hu, F.; Ur-Rehman, M.; von Deneen, K.M.; Tian, J. Wandering Pattern Sensing at S-Band. *IEEE J. Biomed. Health Inform.* **2017**. [CrossRef]

19. Cretikos, M.A.; Bellomo, R.; Hillman, K.; Chen, J.; Finfer, S.; Flabouris, A. Espiratory rate: The neglected vital sign. *Med. J. Aust.* **2008**, *188*, 657–659. [PubMed]

20. Pantelopoulos, A.; Bourbakis, N.G. A survey *on* wearable sensor-based systems for health monitoring and prognosis. *IEEE Trans. Syst. Man Cybern. Part C* **2010**, *40*, 1–12. [CrossRef]

21. Chen, L.; Nugent, C.D.; Wang, H. A knowledge-driven approach to activity recognition in smart homes. *IEEE Trans. Knowl. Data Eng.* **2012**, *24*, 961–974. [CrossRef]

22. Lee, Y.S.; Pathirana, P.N.; Steinfort, C.L.; Caelli, T. Monitoring and analysis of respiratory patterns using microwave doppler radar. *IEEE J. Transl. Eng. Health Med.* **2014**, *2*, 1–12. [CrossRef] [PubMed]

23. Yang, X. Detection of Essential Tremor at the S-Band. *IEEE J. Transl. Eng. Health Med.* **2018**, *6*, 1–7. [CrossRef] [PubMed]

24. Bryan, J.D.; Kwon, J.; Lee, N.; Kim, Y. Application of ultra-wideband radar for classification of human activities. *IET Radar Sonar Navig.* **2012**, *6*, 172–179. [CrossRef]

25. Shany, T.; Redmond, S.J.; Narayanan, M.R.; Lovell, N.H. Sensorsbased wearable systems for monitoring of human movement and falls. *IEEE Sens. J.* **2012**, *12*, 658–670. [CrossRef]

26. Karantonis, D.M.; Narayanan, M.R.; Mathie, M.; Lovell, N.H.; Celler, B.G. Implementation of a real-time human movement classifier using a triaxial accelerometer for ambulatory monitoring. *IEEE Trans. Inf. Technol. Biomed.* **2006**, *10*, 156–167. [CrossRef] [PubMed]

27. Qi, X.; Zhou, G.; Li, Y.; Peng, G. RadioSense: Exploiting wireless communication patterns for body sensor network activity recognition. In Proceedings of the 2012 IEEE 33rd Real-Time Systems Symposium, San Juan, Puerto Rico, 4–7 December 2012; pp. 95–104.

28. Kaushik, A.R.; Lovell, N.H.; Celler, B.G. Evaluation of PIR detector characteristics for monitoring occupancy patterns of elderly people living alone at home. In Proceedings of the 29th Annual International Conference of the IEEE Engineering in Medicine and Biology Society, Lyon, France, 22–26 August 2007; pp. 3802–3805.

29. Zhou, Z.; Chen, X.; Chung, Y.-C.; He, Z.; Han, T.X.; Keller, J.M. Videobased activity monitoring for indoor environments. In Proceedings of the IEEE International Symposium on Circuits and Systems, Taipei, Taiwan, 24–27 May 2009; pp. 1449–1452.

30. Li, C.; Lubecke, V.M.; Boric-Lubecke, O.; Lin, J. A review on recent advances in Doppler radar sensors for noncontact healthcare monitoring. *IEEE Trans. Microw. Theory Tech.* **2013**, *61*, 2046–2060. [CrossRef]

31. Zhang, G.; Yi, T.; Zhang, T.; Cao, L. A multiscale noise tuning stochastic resonance for fault diagnosis in rolling element bearings. *Chin. J. Phys.* **2018**, *56*, 145–157. [CrossRef]

32. Ström, E.G. On 20 MHz channel spacing for V2X communication based on 802.11 OFDM. In Proceedings of the 39th Annual Conference of the IEEE Industrial Electronics Society, Vienna, Austria, 10–13 November 2013; pp. 6891–6896.

33. Halperin, D.; Hu, W.; Sheth, A.; Wetherall, D. Predictable 802.11 packet delivery from wireless channel measurements. In Proceedings of the ACM SIGCOMM 2010 Conference, New Delhi, India, 30 August–3 September 2010; pp. 159–170.

34. Shah, S.A.; Zhang, Z.; Ren, A.; Zhao, N.; Yang, X.; Zhao, W.; Yang, J.; Zhao, J.; Sun, W.; Hao, Y. Buried Object Sensing Considering Curved Pipeline. *IEEE Antennas Wirel. Propag. Lett.* **2017**, *16*, 2771–2775. [CrossRef]

35. Gao, W.; Oh, S.; Viswanath, P. Demystifying Fixed k-Nearest Neighbor Information Estimators. In Proceedings of the 2017 IEEE International Symposium on Information Theory (ISIT), Aachen, Germany, 25–30 June 2017.

36. Sreejith, B.; Verma, A.K.; Srividya, A. Fault diagnosis of rolling element bearing using time-domain features and neural networks. In Proceedings of the IEEE Region 10 and the Third international Conference on Industrial and Information Systems, Kharagpur, India, 8–10 December 2008.

37. Phan, H.; Maaß, M.; Mazur, R.; Mertins, A. Random Regression Forests for Acoustic Event Detection and Classification. *IEEE/ACM Trans. Audio Speech Lang. Process.* **2015**, *23*, 20–31. [CrossRef]

38. Cortes, C.; Vapnik, V. Support-Vector Networks. *Mach. Learn.* **1995**, *20*, 273–297. [CrossRef]

39. Dumais, S.T. Using SVMs for text categorization. *IEEE Intell. Syst.* **1998**, *13*, 21–23.

40. Jerome, F.; Hastie, T.; Tibshirani, R. *The Elements of Statistical Learning*; Springer: New York, NY, USA, 2001.

41. Haykin, S. *Neural Networks: A Comprehensive Foundation*, 2nd ed.; Prentice Hall PTR: Upper Saddle River, NJ, USA, 1994.

42. Rai, V.; Mohanty, A. Bearing fault diagnosis using FFT of intrinsic mode functions in Hilbert-Huang transform. *Mech. Syst. Signal Process.* **2007**, *21*, 2607–2615. [CrossRef]

43. Qin, Y.; Xing, J.; Mao, Y. Weak transient fault feature extraction based on an optimized Morlet wavelet and kurtosis. *Meas. Sci. Technol.* **2016**, *27*, 085003. [CrossRef]

44. Shah, S.A.; Zhao, N.; Ren, A.; Zhang, Z.; Yang, X.; Yang, J.; Zhao, W. Posture Recognition to Prevent Bedsores for Multiple Patients Using Leaking Coaxial Cable. *IEEE Access* **2016**, *4*, 8065–8072. [CrossRef]

45. Breiman, L.; Friedman, J.H.; Stone, C.J.; Olshen, R.A. *Classification and Regression Trees*; CRC Press: Boca Raton, FL, USA, 1998.

46. Kumar, P.S.; Pranavi, S. Performance analysis of machine learning algorithms on diabetes dataset using big data analytics. In Proceedings of the 2017 International Conference on INFOCOM Technologies and Unmanned Systems (Trends and Future Directions) (ICTUS), Dubai, UAE, 18–20 December 2017; pp. 508–513.

47. Wikipedia.org. Available online: https://en.wikipedia.org/wiki/Precision_and_recall (accessed on 12 February 2018).

48. Breiman, L. Random forests. *J. Mach. Learn.* **2001**, *45*, 5–32. [CrossRef]

 *applied sciences*

Article

Emergency-Prioritized Asymmetric Protocol for Improving QoS of Energy-Constraint Wearable Device in Wireless Body Area Networks

Jaeho Lee [1] and Seungku Kim [2,*]

[1] Department of Information and Communications Engineering, Seowon University, 377-3 Musimseoro, Seowon-gu, Cheongju 28674, Chungbuk, Korea; izeho@seowon.ac.kr

[2] School of Electronics Engineering, Chungbuk National University, Chungdae-ro 1, Seowon-Gu, Cheongju 28644, Chungbuk, Korea

* Correspondence: kimsk@cbnu.ac.kr; Tel.: +82-10-9312-9358

Received: 6 December 2017; Accepted: 8 January 2018; Published: 10 January 2018

Abstract: Wireless Body Area Network (WBAN) is usually composed of nodes for contacting the body and coordinator for collecting the body data from the nodes. In this setup, the nodes are under constraint of the energy resource while the coordinator can be recharged and has relatively larger energy resource than the nodes. Therefore, the architecture mechanism of the networks must not allow the nodes to consume much energy. Primarily, Medium Access Control (MAC) protocols should be carefully designed to consider this issue, because the MAC layer has the key of the energy efficiency phenomenon (e.g., idle listening). Under these characteristics, we propose a new MAC protocol to satisfy the higher energy efficiency of nodes than coordinator by designing the asymmetrically energy-balanced model between nodes and coordinator. The proposed scheme loads the unavoidable energy consumption into the coordinator instead of the nodes to extend their lifetime. Additionally, the scheme also provides prioritization for the emergency data transmission with differentiated Quality of Service (QoS). For the evaluations, IEEE 802.15.6 was used for comparison.

Keywords: wireless body area networks; medium access control; energy-balanced model; energy efficiency; Quality of Service

---

1. Introduction

In recent years, Wireless Body Area Networks (WBAN) have been attracting the national interest of biomedical informatics by meeting the communication technology requirement. Many related studies have concentrated on fusing medical services and communication technologies to realize unexplored fields of medical service for enhanced quality of life. This development can conserve the cost of medical services and allow wide distribution of medical knowledge to nonmedical personnel.

In this setup, all information of the users who wear the healthcare equipment should be gathered by micro-sized sensors, which measure and collect body signals, such as electrocardiogram (ECG), electroencephalogram (EEG), or Vital Sign (e.g., heart rate, blood pressure, temperature, pH, respiration, oxygen saturation) [1]. These signals should be delivered to a remote base station for diagnosis and prescription, with harmless personal communication method.

This situation requires all physical and physiologic data monitored from the human body for the synthetic data fusion and the central preprocess with high efficiency of energy utilization. Another significant requirement is guaranteeing the minimized latency for emergency data transmission. It is clear that the emergency data has to be urgently reported as soon as possible for immediate rescue whenever a node senses dangerous signal of body condition.

Meanwhile, communication devices in WBAN are generally composed of nodes sensing the body signal and a coordinator analyzing the sensed body signal. Another significant role of the coordinator is harmonizing multiple nodes against collisions, interferences, and unnecessary energy consumptions. The main difference on the specific environment of WBAN is energy resource. While the nodes contacting the body have extremely limited battery capacity due to their light-weighted and micro-sized feature, a coordinator (e.g., mobile phone) has relatively sufficient energy resource because it has abundant hardware resources and rechargeable battery. Hence, the main focus for designing WBAN should consider the energy balance between nodes and coordinator.

WBAN technology has been investigated for near distance (3 m–10 m) communication within or around the body with conditional radio transmission power which must be ergonomically harmless to human health. Moreover, it is clear that the Medium Access Control (MAC) layer is the most crucial layer in terms of energy efficiency because it handles idle listening, retransmission, carrier sensing, and control of the transmission power. Consequently, WBAN has to meet these restricted requirements and the MAC protocol can become a key technique for addressing these emerging issues.

Furthermore, it is possible to analyze the general conditions of the WBAN environment by focusing on the MAC layer. First, we can find that most of the transmissions might be incurred in uplink direction, that is, from nodes to coordinator. Second, a coordinator should support the registration for newly participating node at any time. Third, all data can be classified into periodic report data and emergency data to allow the MAC protocol to provide different Quality of Service (QoS). Fourth, when the MAC protocol considers energy efficiency, it has to adaptively and asymmetrically control the energy balance between nodes and coordinator.

Many WBAN MAC protocols have long been designed and proposed. However, they have not considered the above issues, especially in adapting different energy balance between nodes and a coordinator, while also guaranteeing the minimized delivery delay for transmission of emergency data. Hence, this motivates us to design a new MAC protocol, called aSymMAC, to analyze the above four issues with dealing solutions on time registration, allocating downlink transmission only if needed, handling QoS between periodic report data and emergency data, and designing an asymmetric energy-balanced model for efficient energy distribution belonging to WBAN environment.

2. Related Work

Given that WBAN was spotlighted, much research has been conducted to address this limited environment. IEEE 802.15.6 [2] standard technology, which was published for communications of WBAN environment in 2012, illustrated an energy-efficient scheme on both Physical (PHY) and MAC layers. This standard technology allowed only the star-topology; therefore, all nodes deployed on the human body should be supervised by a coordinator named as hub, and the scheme on the MAC layer of the standard employed combined approach between slot-assigned and random-accessed methods. The superframe is classified into several time periods and a beacon frame determines the length of each frame duration. In this protocol, all nodes can transmit their reserved data through the slot-assigned period without any contention, denoted as Type-I/II Access Phase. Furthermore, this technology employed Exclusive Access Phase (EAP) and Random Access Phase (RAP) for non-reserved data transmissions. Nodes can use these periods to transmit data when they have emergency data or remaining data in only EAP and RAP with the approach of Carrier Sense Multiple Access/Collision Avoidance (CSMA/CA) or Slotted-Aloha. The main challenge of IEEE 802.15.6 compared to other standard technologies (e.g., 802.15.4 [3]) is the consideration for emergency data transmissions. However, non-reserved emergency data transmission is allowed in only EAP and RAP, so this can be limited point of this protocol.

Besides IEEE 802.15.6, there are many types of research for enhancing and developing the communication method, and H-MAC [4] was a representative novelty mechanism that employs Time-Division Multiple Access (TDMA) approaches for body sensor networks. The general TDMA-based MAC protocols should require base time tick for synchronization such as beacon.

However, H-MAC did simply address this problem by using the heartbeat vibration measured from the physiological signal with some sensors as a synchronization tick for TDMA. Hence, all nodes could transmit data without any contention under the assumption of that all nodes already had the sequence for slot assigned ownership. Nevertheless, they did not receive any beacon frame from the coordinator.

Some research [5,6] had been proposed to facilitate WBAN use-cases with developing previous IEEE 802.15.4 standard which was widely used for wireless sensor networks. Dynamic Time-Division Multiple Access (DTDMA) [7] designed a reservation-based dynamic TDMA MAC protocol for evolving energy efficiency. The overall superframe of this protocol was composed of the beacon, Contention Free Period (CFP), Contention Access Period (CAP), and an inactive period. But the frame sequence of DTDMA structure was contrary to IEEE 802.15.4. This protocol prioritized CFP to minimize the latency for emergency data transmissions, and then assigned CAP for periodic or remaining data transmissions. The rest of the technical parts including the beacon interval and the length of the inactive period complied with IEEE 802.15.4 protocol.

Another representative scheme of the 802.15.4-based protocol is BodyMAC [8]. In this protocol, device roles were classified into coordinator and implant, and these would be matched to the hub and node in IEEE 802.15.6. This protocol had cut off the inactive period from IEEE 802.15.4 and newly assigned downlink subframe instead of CFP in the superframe. All implants are guaranteed to be assigned to respective downlink slots to conserve energy from idle. Furthermore, this protocol allowed that some implants do not receive a beacon from the coordinator. The coordinator will perform wake-up signal transmission process to activate a corresponding implant if it had data to be sent. After the sleeping implant was activated with this process, it synchronizes to the wake-up signal and receives data from the coordinator after finding the next beacon frame.

More recent research for WBAN has been also conducted with the same goal. Moulik et al. [5] designed a priority-based MAC frame depending on the energy efficiency of the healthcare system. This research employed Fuzzy interference system to accomplish an optimization method for individual sensor nodes which can have independent environments, such as radio status, data length, or transmission interval derived from divergent sensor types. The main contribution of this work is to consider the different characteristics from individual node to find the optimization of ideal energy efficiency.

In addition, Zhou et al. [6] developed an intelligent management method for WBAN with game theory. The authors endeavored to find an appropriate way in MAC layer between contention access approach and contention free approach. In this work, a network has dynamic strategies considering the application requirements and radio channel status so that the proposed scheme supports flexible energy resource allocation. This research also has an optimized method for energy efficiency to satisfy the requirements from various networks. Other researches tried to consider WBAN routing environment [9] and to design low power MAC [10].

3. Protocol Framework

Recently, many types of research related to WBAN has been designed to handle the energy efficiency and latency under the human body network environment and they have made many contributions. However, some points to satisfy the significant requirements of WBAN remain as challenging issues and these could be classified into two phenomena below.

The first point is the immediate transmission for emergency data. Most research, except for the IEEE 802.15.6 standard, focused on the overall energy efficiency and overall latency, but could not make an effort for emergency data transmission because they just allocate separate time slots for this purpose. Moreover, it is not clear that the standard technology of IEEE 802.15.6 efficiently accomplished this issue under EAP and RAP; it just employed the Slotted Aloha and CSMA approach without any prioritization for emergency data. Specifically, all data were just fairly transmitted during these periods regardless of whether it is an emergency or not. The emergency data can be incurred at any time regardless of time slot structure. Therefore, the superframe-based WBAN MAC protocol has to support an immediate transmission method for emergency data within the overall time domain.

The second fundamental requirement for WBAN is that the energy balance between nodes and coordinator must be differently considered. Most smart devices (e.g., phone, watch, and wristband) can generally play a role of coordinator in WBAN environment because they have more sufficient energy resource than the nodes, and their battery might be rechargeable. WBAN MAC protocols have to consider this issue by asymmetrically allocating the energy consumption point on the both roles.

As a result, the proposed scheme on this paper was focused on the above two significant issues to satisfy the essential requirement of WBAN environment by handling the emergency data transmissions and designing asymmetric energy balance model between nodes and coordinator.

IEEE 802.15.6 standard technology employed CSMA-based mechanism when transmitting emergency data from nodes. However, CSMA-based mechanism absolutely depended on Clear Channel Assessment (CCA) which was provided in a radio transceiver, but the reliability of CCA effect cannot be guaranteed because Medical Implant Communication Service (MICS) band, which is a widely used radio band for WBAN, does not permit over the transmission power of -85 dBm. Moreover, CSMA-based protocols have some energy-waste problems (e.g., idle listening, overhearing, protocol overhead, and preamble sampling). With these reasons, we found that CSMA-based protocol is not appropriate for WBAN and decided to hybridize both CSMA and TDMA for WBAN MAC protocol.

Before describing the proposed protocol in detail, we named the coordinator as Personal Coordinator (PC) and assumed that the communications of node-to-node were not presented because general WBAN nodes report their sensed data only to a coordinator. Note that this paper is mainly aiming to design of both asymmetrical energy-balanced model and QoS for emergency data transmission. Based on the prioritized scheme, the proposed protocol concurrently supports both transmission types of normal and emergency without any operation mode change, contrarily to IEEE 802.15.6.

3.1. Basic Framework

Figure 1 illustrates the overall structure of the MAC protocol which we proposed in this paper. The superframe is composed of a Beacon frame and multiple subframes, and each subframe is combined with an Active frame and a Sleep frame. Furthermore, each Active frame at every subframe is composed of several periods; Sync, Transmission Period (TP), and Contention Period (CP) as shown in Figure 2. TP in the Active frame is composed of both d-TP and u-TP; d-TP denoted for downlink directional transmission period from a PC to an individual node, and u-TP denoted for uplink transmission.

The beacon packet is generated from PC and it is disseminated to all nodes in Beacon frame with relatively long term interval denoted as T_B. As discussed in the previous paragraph, one beacon frame is followed by multiple frames in each T_B as shown in Figure 1. Moreover, T_S, which denotes the time duration of each subframe, has relatively short term value compared with T_B. Each subframe is composed of an Active frame and a Sleep frame in which the RF transceiver should keep the wake-up condition and the sleep condition. Hence, T_B could be considered as the duty cycle of the transceiver. Moreover, in Sleep frame, the PC and all nodes enter to the condition of deep sleep or standby status for energy conservation. Hence, all transmissions are basically absent during Sleep frame, but we allowed for an exception in the case of emergency data transmission.

To synchronize every frame between the coordinator and each correspondent node, the PC should periodically transmit the beacon packet containing both T_B and T_S. Every node belonging to the same coordinator can be aware of the start time to be awake depending on T_B and reception time of the packet, and each node can also estimate the reception time of the next beacon packet from the same PC. Consequently, if a node does not have any packet to send, it can omit all forwarding frames and can just sleep until the next Beacon frame.

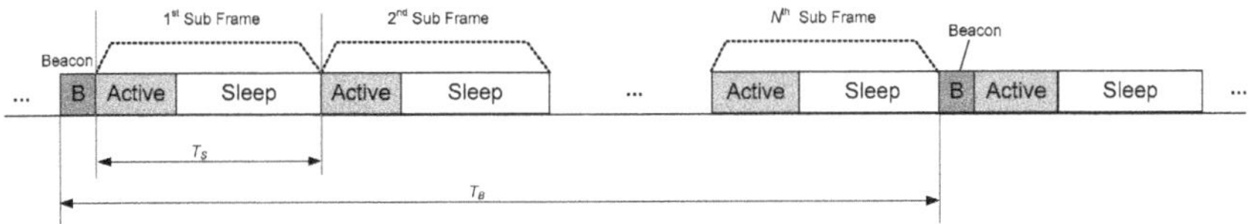

Figure 1. Overall frame structure of the proposed aSymMAC protocol.

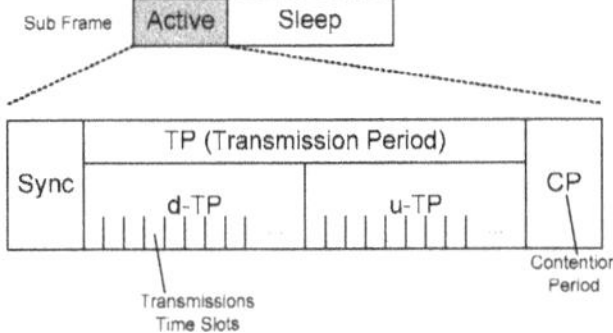

Figure 2. The 3-period types (Sync, Transmission, and Contention Period) in the Active subframe.

On the other hand, Sync period is for disseminating Sync packet to awaken nodes from PC, which contains each time offset values for d-TP, u-TP, CP, and Sleep frame. The Sync packet is always sent from a PC at every T_B, and we design that each subframe can be alternatively applied to each node. Each corresponding node should decide to participate in the current subframe or not. The rest of a subframe is CP, and we adapt the exceptional case of allowing contentions based on CSMA/CA from nodes on this period, in which this period supports the registration of a new node joining in.

Under the given timeframes above, all transmissions except in CP should be critically controlled by the PC. By the way, emergency data transmission must be transmitted without any authentication by PC. IEEE 802.15.6 protocol allows the emergency data transmission by out of coordinator, but it allowed in only EAP and RAP. However, emergency data related to the biomedical signal must be immediately sent regardless of the timeframe condition. For this reason, we also designed a mechanism to minimize the latency of emergency data transmission. In the proposed scheme, the emergency data transmission can be allowed in overall time frames (i.e., d-TP, u-TP, CP. And even Sleep frame) with prioritized channel preemption.

As we described previously, the PC relatively has sufficient energy resource than nodes so that the main goal of the proposed MAC protocol is designing an asymmetrical energy balance model between PC and nodes. Meanwhile, it is fundamentally required that the energy consumption point should be loaded into a PC as possible for conserving the energy of nodes side. The nodes should wake up on the only essentially required moment.

3.2. Transmit Period (TP)

TP is assigned for transmission duration between PC and nodes. Under the characteristics of WBAN, we focused on transmission topology as Piconet cluster which is used in Bluetooth technology. The direct transmission between nodes is not allowed, but can be forwarded via PC. Hence, we considered two-way transmissions of downlink and uplink in TP.

d-TP allows data to be sent from PC; both unicast and broadcast methods. Regardless of unicast or broadcast, the PC can transmit data anywhere, but the nodes cannot be aware of when the PC can send to them. Hence, we divide the d-TP by multiple time slots, and considered both unicast and broadcast. The Sync packet contains time offsets, destination address, and type (unicast or broadcast), depending on each d-TP slot.

After a node receives the Sync packet from the PC, it finds all d-TP slots assigned to broadcast type, and also finds the destination addresses matched to itself if the type is unicast. Based on the Sync packet, each node can be aware of when it has to be awake to receive data from PC and when it can sleep.

If a node is selected as a receiver but cannot detect any preamble after Inter Frame Space (IFS), it should process an Early-sleep mechanism [11]. Note that IFS is existed for guaranteeing the turnaround time of transceiver transaction and for supporting the prioritized emergency data transmission.

For the next step, u-TP is also divided into multiple time slots as shown in Figure 2, and all slots are fairly assigned for all nodes. In this period, all nodes do not have to keep a wake-up state. The node only wakes up if it has data to send to the PC. After a node finds the schedule of d-TP from the Sync packet reception, it accesses the assigned d-TP slots and transmits data. If a node cannot complete the transmission during the assigned time slots, it can continue to transmit the remaining data during CP using the competition approach. Moreover, the PC always wakes up during overall u-TP for successful reception.

In both d-TP and u-TP, the Ack mechanism should be performed within the same time slot of the corresponding data transmission as shown in Figure 3. Specifically, one-time slot covers one data packet and one Ack packet. That is, the overall TP can be considered as a contentionless period because any packet does not collide in this period.

The downlink transmission occurs only when the PC will configure or change some parameters of time schedule. u-TP is helpfully used to report periodic medical signals, such as vital sign, ECG, and EEG, meanwhile d-TP is not frequently required. For these characteristics, we employed selective d-TP assignment to allow the Sync packet to include the information of d-TP schedule only when it is required.

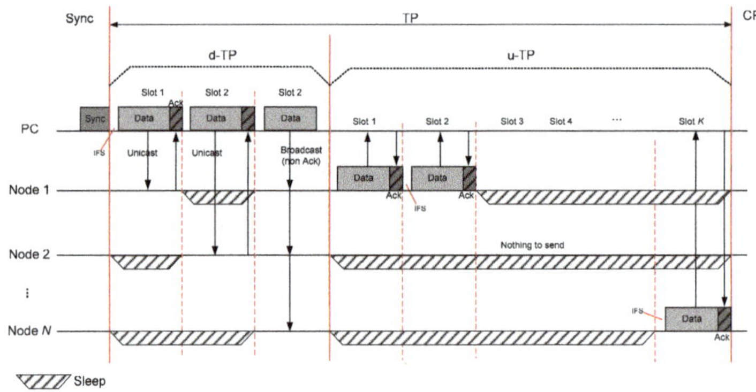

Figure 3. Transmission example of TP (Transmission Period) which composed of d-TP (downlink-TP) and u-TP (uplink-TP) in the Active subframe.

3.3. Contention Period

CP is assigned for transmission of remaining data to be sent to the PC from the node and for transmission of specific control packets such as new node registration or slot request for u-TP. In this period, as shown in Figure 4, all nodes and PC are performing transmission including emergency data with CSMA/CA mechanism after IFS duration since the start time of CP.

All transmissions in CP can be classified into the remaining data, emergency data, and specific control frames such as registration and u-TP slot request. If a node or some nodes did not complete data transmission during the assigned u-TP, they can additionally continue to transmit it in this period with competition based on CSMA/CA, so this period can be alternatively used with TP. Furthermore, if a node tries to register at the given network or changes its slot, it can also try to request to the PC via this period.

As the mechanism of TP, CP also employs Ack so all types of packet should be followed by Ack. After PC receives a request for registration or slot change, it should apply and broadcast the changed schedule after the next subframe through the Sync packet. On the contrary, PC can deny the request from nodes by notifying Nack to the corresponding node if there is any reason to reject the request (e.g., overheated congestion or overflowed capacity). In this case, the node neglected by PC can try to again at the next Beacon time. Note that all transmissions should be competitively sent based on CSMA/CA regardless of any direction (downlink or uplink), and also note that PC should also participate in the competitive transmission with the nodes in CP.

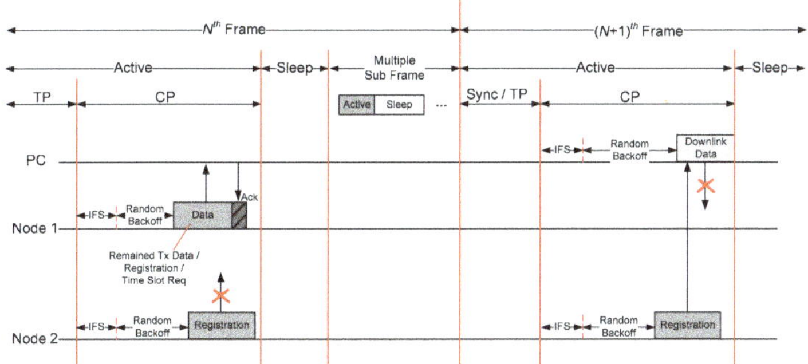

Figure 4. Transmission example of CP (Contention Period) in the Active subframe.

Basically, CP would be required in a few cases. Hence, it is very low frequency to be incurred. To address this situation, it is recommended to allocate CP whenever multiple frames are passed. As a result, all nodes can participate in the competition of CP when it is detected via Sync packet. If a new node failed to be registered, it should keep going on sleep state before the next CP which can be found by each Sync packet.

3.4. Sleep Frame

Presently, many low-powered MAC protocols have employed the Duty Cycle mechanism to overcome the energy constraints so this paper also employed it. As described in Basic Framework subsection, every subframe is composed of an Active frame and Sleep frame. There is no transmission in Sleep frame except emergency data transmission, so all nodes maintain the sleep state during Overall Sleep frame. However, the PC has to perform the Duty Cycle mechanism by periodically switching both wake-up and sleep status because the proposed MAC protocol adapts emergency data transmission with conservation of energy efficiency at most periods.

4. Emergency Data Handling

In the Healthcare system, it is fundamentally required to provide minimized latency for immediate transmission of emergency data when the human body is suffering from severe emergency situation. However, many types of research have just focused on separated time slot assignment for emergency data transmission, and they could not fulfill the optimized solution due to holding contention-free approach. Hence, we have focused on finding the appropriate solution to guarantee prioritized QoS of emergency data transmission. In the proposed MAC protocol, the emergency data can be sent at any time.

4.1. IFS for Active Slot Period

Figure 5 shows the sample case where a node transmits its emergency data during d-TP and u-TP. As previous section including Figures 3 and 4, all data have to be delayed for a short IFS time before transmission according to the overall timeframe (d-TP, u-TP, and CP), except the Sync period. Thus, it clearly provided QoS because the emergency data could be sent by preempting the time moment of IFS in overall TP and CP, prior to other transmissions.

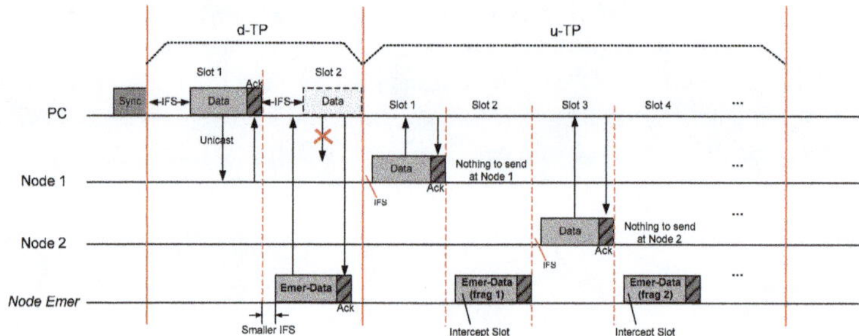

Figure 5. Emergency transmission with prioritized IFS in d-TP and with idle slot preemption in u-TP.

In the left side of Figure 5, the emergency data transmission of Node_Emer can be earlier started with short IFS (SIFS) than PC during IFS in d-TP, so the emergency data can be prioritized to be transmitted even though the time slot was not assigned to Node_Emer. In this case, the PC should perform CSMA/CA to guarantee the reliability and QoS of the emergency transmission if it is existed. This process could be identically operated in CP. Although the nodes should consume energy due to carrier sense in CP, this period is seldom allocated in terms of overall superframe.

Meanwhile, the right side of Figure 5 shows how the emergency data could be transmitted within u-TP. Contrary to the case of d-TP, general data transmission in u-TP is relatively significant because this period is the main stem in the healthcare system. Hence, the proposed MAC protocol allowed for the emergency data to be sent only when the idle slot is detected, for conserving energy of the nodes from carrier sense. We believe that the average delay of emergency data transmissions is not critical because not all nodes are using all assigned slots.

Even though TP (including d-TP and u-TP) and CP are existed for normal data and special control packet transmissions, the emergency data can be concurrently transmitted with high priority with maintaining normal data transmission function. The only weak point is that a node consecutively transmits massive emergency data. However, this situation could be easily addressed by conducting fragmentation if the required time is greater than a time slot, as shown in the right side of Figure 5. Furthermore, we considered that PC should accept an emergency data from a node even though it was not registered to PC before.

4.2. Energy-Detection Method for Emergency Transmission on Sleep Frame

As described in previous section, while the nodes sleep in Sleep frame, the PC performs Duty Cycle during the same time for receiving the emergency data transmissions from any node, regardless of whether it is registered or not. During the Sleep frame, we employed LPL (Low Power Listening) mechanism on PC derived from B-MAC [12]. In this frame, while the nodes must not transmit any general data, the PC has to perform Duty Cycle to detect the preamble signal of the emergency transmission in Sleep Frame and will keep the wake-up state if any preamble is detected because every node can transmit emergency data at any time in this frame.

If a node tries to transmit emergency data in this frame, it configures larger preamble length than the sleep period of the Duty Cycle as shown in Figure 6, to prevent the PC from missing the emergency data due to Duty Cycle. In this situation, Ack is not required because the PC has been absolutely awake since it has detected preamble signal from the node. The nodes are allowed to send data only in case of emergency condition during this frame. Consequently, the PC can successfully receive the emergency data transmissions as well as maintain the energy efficiency.

As a result, since the length of Sleep frame is initially configured, PC performs duty cycle based on LPL mechanism, and nodes just constantly sleep in this period if they don't have emergency data transmission. If a node has emergency data to be transmitted, it just sends data with a long preamble signal.

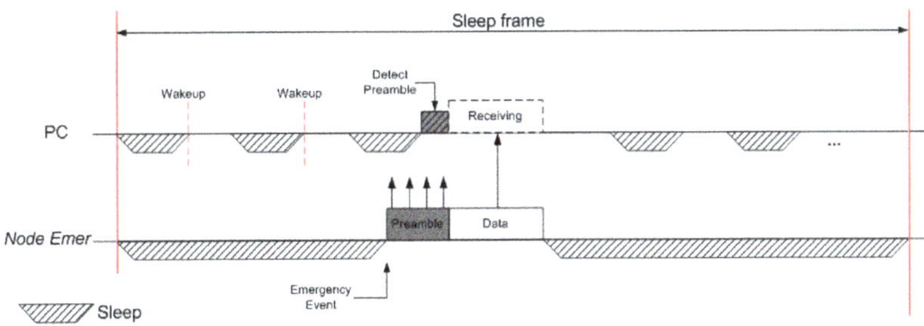

Figure 6. Duty cycle mechanism for emergency data reception in Sleep subframe from Coordinator.

5. Performance Evaluation

To achieve an appropriate evaluation for the proposed MAC protocol, we have evaluated it with two approaches both numerical analysis and experiments. The aSymMAC has not only shown a reasonable superframe structure to allow periodically reporting normal data measured by on-body nodes such as Vital Sign and ECG, but also outperforming the method that can concurrently achieve minimized delivery delay for emergency data transmissions during normal data handling, as described in previous section. From here on, we would describe and prove the performance of our proposed scheme in terms of transmission delay and energy efficiency, for both normal and emergency data transmissions.

5.1. Analytic Definitions and Experimental Environment

Table 1 shows symbol definitions for further convenience in explaining the numerical analysis, which would be illustrated in the next section. For clarification, we assume that all data have same size for protocol utilization.

Meanwhile, we have performed experiments with the simulation environment presented in Table 2. For simulating a real environment, we consider many variables of data length with MTU (Maximum Transmission Unit) and power consumption in each state (Tx, Rx, Sleep), referring data sheets of CCxx families from TI Inc. [13], which are popularly used for low power communications. IEEE 802.15.6 standard technology was employed for the comparison results.

Table 1. Symbol descriptions for analyses.

Symbol	Description	Symbol	Description
N	Number of Nodes	E^C, E^N	Energy Consumption of respective PC and Node
k	Number of Time Slots	T_{IFS}, T_{Data}	Time delay for IFS and data transmission
η	Number of Active Frames	R	Data Rate
Γ	Number of Remained Data	L^n, L^e	Average Latency of respective normal and emergency data transmissions
π	Number of Data Tx Event on a Node during u-TP	P	Power (Watt) value
$\varepsilon_n, \varepsilon_c$	Clock Drift at Nodes and PC	D	Duration of Time Frame
ω	Packet Length of Constant Data	Φ	Duration of Long Preamble Transmission for LPL during Sleep Period
λ	Occurrence Rate of Normal Data Transmission	O	Delay of Pseudo-random back-off

Table 2. Simulation parameters.

Layer	Parameters	Value
PHY	Data Rate	64/256 kbps
	Tx Power	2.428 (0 dBm) mW
	Rx Power	1.814 mW
	Sleep Power	0.027 mW
MAC	Number of PC	1
	Number of nodes	1 to 5
	Topology	Star
	Traffic Pattern	128 byte/Given Interval
	Communication Range	5 to 10 m
	Beacon Interval	1 s
	Period of a Sub-Frame	20 ms
	Active/Sleep ratio	1:4 (200 ms:800 ms)
	Duration of u-TP	120 ms
	Duration of d-TP	40 ms if only required
	Duration of CP	40 ms
	Number of time slots in u-TP	6
	Number of time slots in d-TP	2
	Number of Active frames	10
	Duty cycle interval of PC	1 ms
	Length of long preamble	950 µs
	Clock drift for PC and nodes	40 ppm
	Occurrence of control frame in CP	less than 1%
	Length of IFS	150 µs
	Length of SIFS	50 µs

5.2. Latency Analysis of Emergency Transmission

Meanwhile, the transmission latency of normal data does not need minimized delay, emergency data must be delivered to PC as immediate as possible. For the evaluations, we analyzed the numerical results of the transmission delay from both normal data and emergency data transmissions.

The overall time duration at the viewpoint of the whole superframe expressed as D_{super} could be presented as Equation (1).

$$\begin{aligned} D_{super} &= \left(D_{active} + D_{sleep} \right)\eta + D_{beacon} \\ &= \left(D_{sync} + D_{d-TP} + D_{u-TP} + D_{CP} + D_{sleep} \right)k + D_{beacon} \end{aligned} \tag{1}$$

defined by the following equation. The opportunity of nodes to send normal data should exist in u-TP and CP. Hence, the equation employed L variable in these periods, compared with the constant time duration denoted as D.

$$L_{total}^{n} = \frac{1}{2}\left(L_{u-TP}^{n} + L_{CP}^{n} + D_{sync} + D_{d-TP} + D_{sleep} + \frac{D_{beacon}}{\eta} \right) \tag{2}$$

Moreover, at the viewpoint of the respective time period, the latency of $L_{u\text{-}TP}$ at a given node denoted as m could be described as Equation (3), considering that the probability of opportunity to transmit at kth time slot might be k/n.

$$L_{u-TP(m)}^{n} = \frac{k_u}{n}\left(T_{IFS} + T_{Data} + T_{Ack} \right) \tag{3}$$

Hence, the average latency of each node during both u-TP and CP denoted as respective $L_{u\text{-}TP}$ and L_{CP} could be presented as follows, where the deviation of time drift of nodes and PC are denoted as ε_n and ε_c, respectively. In Equation (5), we needed to consider the occurrence ratio of each control frame in CP, such as slot request or registration request denoted as λ_{slot} and λ_{reg} variables, respectively.

$$
\begin{aligned}
L_{u-TP}^n &= \frac{k_u}{n^2}\sum_{\delta}^{n}(T_{IFS} + T_{Data} + T_{Ack}) \cdot 2(\varepsilon_n + \varepsilon_c) \cdot \delta \\
&= n^{-1}k_u(n+1)(T_{IFS} + T_{Data} + T_{Ack})(\varepsilon_n + \varepsilon_c)
\end{aligned}
\tag{4}
$$

$$
\begin{aligned}
L_{CP}^n &= O_{avr} + \left(1 - \left(\gamma + \lambda_{reg} + \lambda_{slot}\right)^{-1}\right)(T_{Data} + T_{Ack}) \\
&= T_{IFS} + \tfrac{1}{2}(O_{MAX} + O_{min}) + \left(1 - \left(\gamma + \lambda_{reg} + \lambda_{slot}\right)^{-1}\right)(T_{Data} + T_{Ack})
\end{aligned}
\tag{5}
$$

Meanwhile, the emergency data could be transmitted in d-TP, u-TP, CP, and Sleep frame if each condition is satisfied, so the expected latency of the emergency data transmission should be differently described as below Equation (6) by Equation (1).

$$
L_{total}^e = \frac{1}{2}\left(L_{d-TP}^e + L_{u-TP}^e + D_{sync} + L_{CP}^e + L_{sleep}^e + \frac{D_{beacon}}{\eta}\right)
\tag{6}
$$

Contrary to normal data transmission, emergency data can be sent at any time if allowed. Since it has smaller IFS denoted as T_{SIFS} than the other case(T_{IFS}) to preempt the opportunity to transmit with high priority in d-TP and CP, the emergency data should be accepted to transmit only when the condition of the given time slot is idle in u-TP. Hence, in each time slot, we have considered the number of nodes and have seen the probability of whether other normal data transmission is incurred or not, denoted as ρ. By these cases, the expected value of each period could be presented as follows.

$$
L_{d-TP}^e = T_{SIFS} + T_{Data} + T_{Ack}
\tag{7}
$$

$$
\begin{aligned}
L_{u-TP}^e &= (\rho_1 + \rho_2 + \rho_3 \cdots \rho_{k_u})(T_{IFS} + T_{Data} + T_{Ack}) \\
&= \sum_{\delta}^{k_u}\left\{1 - (1 - \lambda_u)^n\right\}^{\delta-1} \cdot (1 - \lambda_u)(T_{IFS} + T_{Data} + T_{Ack}) \\
&= \left[1 - \left\{1 - (1 - \lambda_u)^n\right\}^{k_u}\right](T_{IFS} + T_{Data} + T_{Ack})
\end{aligned}
\tag{8}
$$

As mentioned in Section 4.2, the PC just performs the duty cycle mechanism in Sleep frame, and the node which has emergency data has to transmit a fixed long preamble that should be larger than the interval of the duty cycle that the PC performs. In the nodes, the latency in CP has the same value in d-TP because the nodes get always prioritized transmission opportunity in d-TP, compared with PC. Therefore, the latency of emergency data transmission in Sleep frame could be the following Equation (9) and the latency of that in CP also could be Equation (10).

$$
L_{sleep}^e = \varphi + T_{Data}
\tag{9}
$$

$$
L_{CP}^e = L_{d-TP}^e
\tag{10}
$$

With the above equations from Equations (1)–(10), we performed an experiment with the environmental parameters shown in Table 2. We found that the latency of emergency data transmission showed outstanding results compared with the normal data transmission. In this experiment, we employed one coordinator and variable nodes. Figures 7 and 8 show the comparison results according to the number of nodes and traffic amount in each transmission rate.

Figure 7 shows the results of average latency according to the number of nodes where they have same traffic. In this case, we fixed the traffic amount from all nodes as 1 packet/s per node and increased the number of the node to verify the latency of normal data transmission. Figure 8 shows the average latency from all nodes according to the variable traffic. Even though the latency of normal data transmission could not show the enhanced latency compared with IEEE 802.15.6, it could not say that the proposed MAC has more weakness because the amount of differentiation is too small and this could be changed by adjusting the parameters related to time schedule. Moreover, we could find that aSymMAC showed better performance than IEEE 802.15.6 beyond 0.4 of traffic interval in Figure 8. Therefore, the proposed scheme would show enhanced latency at high traffic environment.

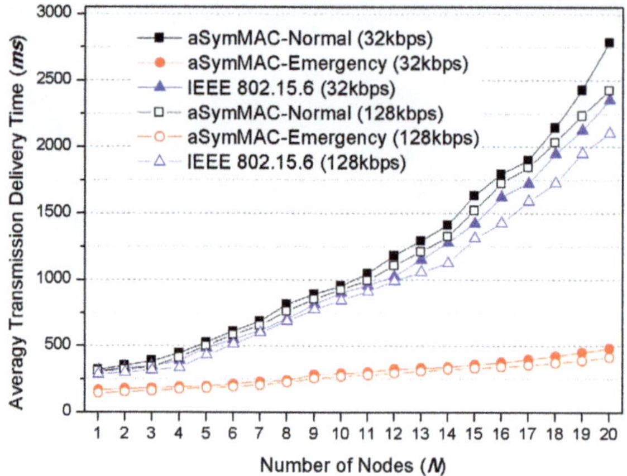

Figure 7. Average delivery time of normal and emergency data transmissions according to the increment of the number of nodes compared with the IEEE 802.15.6 protocol.

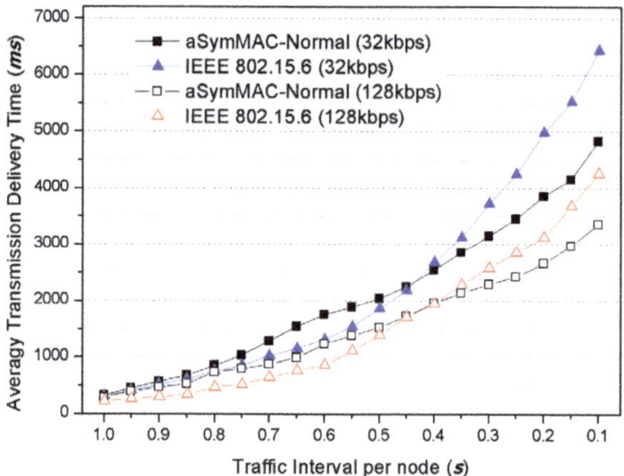

Figure 8. Average delivery time of normal data transmissions according to the traffic amount compared with the IEEE 802.15.6 protocol.

To consider the different speed on the PHY layer, we simulate 128 kbps and 32 kbps of transmission rates and present the results in Figures 7 and 8. Both results similarly show the pattern of differentiation of delivery time on IEEE 802.15.6 and normal transmission on the proposed scheme; higher transmission rate induces lower delivery time. However, it cannot easily say that the differentiation of two transmission rates is high, because both mechanisms are depending on slot assignment structure. Meanwhile, in high traffic environment, high transmission rate can mitigate congested competition during contention period (RAP in IEEE 802.15.6 and CP in aSymMAC). Moreover, this condition can provide more opportunities to transmit normal data during contention-less period (Type I/II in IEEE 802.15.6 and u-TP in aSymMAC). Hence, Figure 8 shows the higher differentiation results of both protocols in high traffic environment.

Contrary to Figures 7 and 8, the latency of the emergency data transmissions show different results as illustrated in Figure 9. In this experiment, we constantly fixed the number of nodes as 4 and also fixed the normal data transmission amount on each node as shown in Table 2. Then, we set another node to transmit only emergency data with varied traffic amount. We finally measured outperforming results comparing with any other scheme as shown in Figure 9.

The differentiation of each transmission rate in the results of IEEE 802.15.6 protocol can be explained with the same reason described in Figures 8 and 9. However, we have to focus on the emergency data transmission on aSymMAC. In this protocol, every emergency data can obtain high opportunity to be transmitted by prioritized IFS (i.e., SIFS), preliminarily to normal data that has relatively longer IFS. Consequently, Figure 9 shows narrow differentiation results of this case.

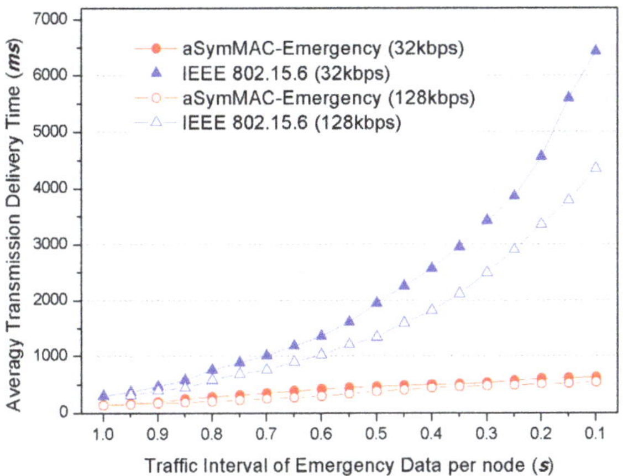

Figure 9. Average delivery time of emergency data transmissions according to the traffic amount with fixed normal data transmissions compared with IEEE 802.15.6 protocol.

Figures 10 and 11 show the saturated throughput with various payload length and the amount of external interference. The saturated throughput presents the relative value of bandwidth efficiency where the maximized value (i.e., 1.0) implies that the given channel is being constantly used for data transmission at every time. Hence, the value of 1.0 is ideal and unrealistic because some overheads are essentially required (e.g., packet header, Ack, IFS, Back-off, etc.) to be operated by MAC protocol.

In the results of IEEE 802.15.6 and normal data transmission of aSymMAC in Figure 10, both protocols similarly present that less payload length makes lower bandwidth efficiency due to the heavy overhead; header and Ack mechanism are commonly applied even if data payload is too small. On the other side, we can see that the excessive payload length makes worse performance caused by that too larger payload length overflows assigned time slot and aggravates competition on contention period. Moreover, this can make more collisions. Meanwhile, the case of emergency transmission on aSymMAC shows better performance because every emergency transmission can preliminarily preempt the channel regardless of slot assignment.

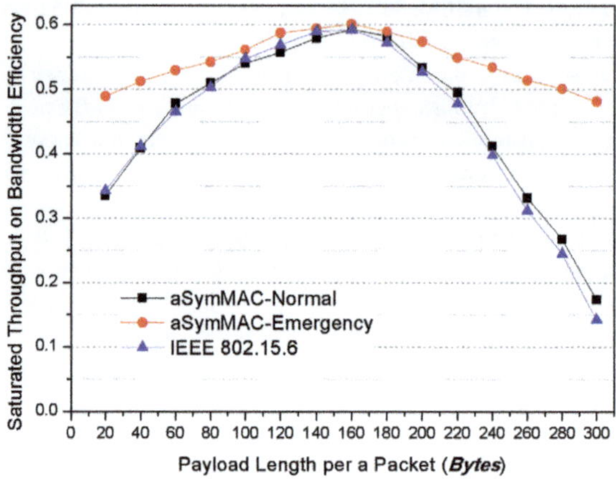

Figure 10. Average saturated throughput depending on bandwidth efficiency according to the payload length per a packet compared with IEEE 802.15.6 protocol.

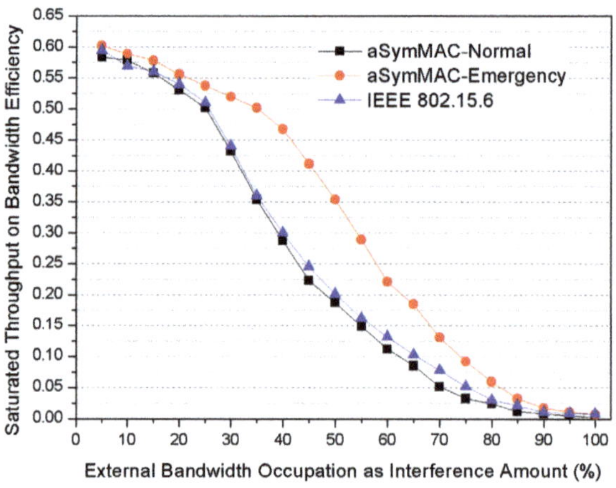

Figure 11. Average saturated throughput depending on bandwidth efficiency according to the external interference on the same frequency compared with IEEE 802.15.6 protocol.

In addition, we evaluate the robustness of these protocols by giving external interference on the same frequency. We employed IEEE 802.15.4 as an interference method with the change of frequency and channel space. Note that the value of 100% implies that IEEE 802.15.4 fully occupies the given channel in time domain. All results at 100% of the interference are not equal to zero because IEEE 802.15.4 also performs CSMA/CA during contention period. As a result, we can see the better performance from the case of emergency transmission on aSymMAC. The emergency transmission can take more opportunities due to cutting off CSMA/CA while IEEE 802.15.4 does the opposite.

5.3. Energy Efficiency of Nodes and PC

In this subsection, we described the results of energy efficiency of the proposed MAC by analyzing the energy consumption from both PC and nodes. Depending on the parameters described in Table 1, the overall energy consumption of PC denoted as E^C could be illustrated as follows:

$$
\begin{aligned}
E^C &= E^C_{active} + E^C_{sleep} + \frac{E^C_{beacon}}{\eta} \\
&= E^C_{sync} + E^C_{d-TP} + E^C_{u-TP} + E^C_{CP} + E^C_{sleep} + \eta^{-1}E^C_{beacon}
\end{aligned}
\tag{11}
$$

The energy consumption of Sync and Beacon periods within one superframe could be simply described as below equations where P parameter would be classified into P_{Tx}, P_{Rx}, and P_{idle} on each radio transceiver condition (Tx, Rx, and idle listening status). Note that the duration of each period is not equal to the interval value.

$$
\begin{aligned}
E^C_{sync} &= P_{Tx}D_{sync} \\
E^C_{beacon} &= P_{Tx}D_{beacon}
\end{aligned}
\tag{12}
$$

In TP, the data transmission from a PC can be incurred or not. Therefore, the analysis of energy consumption in u-TP and d-TP has to consider the occurrence probability of data transmission event and dependently assigned uplink and downlink time slot numbers denoted as k_u and k_d, as below Equations (13) and (14), respectively.

$$
\begin{aligned}
E^C_{u-TP} &= (P_{idle}T_{IFS} + P_{Rx}T_{Data} + P_{Tx}T_{Ack})\lambda_u k_u \\
&\quad + P_{idle}(T_{Data} + T_{Ack})(1 - \lambda_u)k_u \\
&\quad + (P_{idle}T_{IFS} + P_{Rx}T_{Data} + P_{Tx}T_{Ack})\lambda_e k_u \\
&= k_u(\lambda_u + \lambda_e) \\
&\quad \times \left(P_{idle}\left(T_{IFS} + \left((\lambda_u + \lambda_e)^{-1} - 1\right)(T_{Data} + T_{Ack})\right) + P_{Rx}T_{Data} + P_{Tx}T_{Ack}\right)
\end{aligned}
\tag{13}
$$

$$
\begin{aligned}
E^C_{d-TP} &= k_d(P_{idle}(T_{IFS} + (1 - \lambda_d)(T_{Data} + T_{Ack})) \\
&\quad + P_{Tx}(\lambda_d T_{Data} + \lambda_e T_{Ack}) + P_{Rx}(\lambda_d T_{Ack} + \lambda_e T_{Data}))
\end{aligned}
\tag{14}
$$

In addition, the overall energy consumption of the PC in CP could be illustrated as below equation where both the energy consumptions for emergency data and normal data transmission were denoted as $E_{Emergency}$ and E_{Normal}, respectively.

$$
E^C_{CP} = \lambda_e E_{Emergency} + (1 - \lambda_e)\left(\lambda_{reg} + \lambda_{slot} + \lambda_{remain}\right)E_{Normal}
\tag{15}
$$

For more simplification of the analysis, we solve the above equation where the length values of all packets including the slot request, registration request, and request for remaining data transmission have the same size.

$$
\begin{aligned}
E^C_{CP} &= \lambda_e(P_{Rx}T_{Data} + P_{Tx}T_{Ack}) \\
&\quad + (1 - \lambda_e)\left((\lambda_{reg} + \lambda_{slot} + \lambda_{remain})(P_{Rx}T_{Data} - P_{idle}(T_{Data} + T_{Ack})) + \lambda_{remain}P_{Tx}T_{Ack} + P_{idle}(T_{Data} + T_{Ack})\right)
\end{aligned}
\tag{16}
$$

In sleep period, a PC should perform LPL mechanism for allowing data reception of emergency data from any node. Hence, we could describe the energy consumption of a PC with the values of duty cycle interval, wake-up duration, and long preamble length depending on LPL respectively denoted as T_{duty}, T_{wakeup}, and ω_p, as below equation.

$$
E^C_{Sleep} = P_{Rx}\lambda_e\left(\frac{\omega_p}{2R} + T_{Data}\right) + (1 - \lambda_e)P_{idle}D_{Sleep}\frac{T_{wakeup}}{T_{duty}}
\tag{17}
$$

Meanwhile, the average energy consumption value of nodes could be also described as below.

$$
E^N = E^N_{sync} + E^N_{d-TP} + E^N_{u-TP} + E^N_{CP} + E^N_{sleep} + \eta^{-1}E^N_{beacon}
\tag{18}
$$

The energy consumption of a node could be induced by the same approach from Equation (12) with just cross match between the transmitter and receiver, as below equation.

$$E_{sync}^{N} = P_{Rx}D_{sync}$$
$$E_{beacon}^{N} = P_{Rx}D_{beacon}$$

$$(19)$$

In this case, a node may receive only one packet or nothing from a PC in d-TP, so the energy consumption of a PC in this period could be presented as below.

$$E_{d-TP}^{N} = \lambda_d(P_{Tx}T_{ACK} + P_{Rx}T_{Data}) \\ + P_{idle}(\lambda_d T_{IFS} + (\lambda_d + k_d - 2\lambda_d k_d)(T_{Data} + T_{ACK}))$$

$$(20)$$

In u-TP, all nodes possibly have variable transmission event to the corresponding PC. The probability of an uplink event from a node can be affected by the cumulated transmissions that were not sent during the last u-TP and the last CP. Additionally, the number of uplink transmissions from a node could be limited by the number of assigned time slot. For simplification, the analysis of the energy consumption could be calculated under the assumptions that all slots are fairly distributed to all nodes but that all nodes have different number of uplink transmissions, as below equation. Note that the maximum slot number per a node could be presented as $[K_u/N]$ with the Gauss function, with the number of data presented as π.

$$E_{u-TP}^{N} = \sum_{l=0}^{[\frac{K_u}{N}]} P(\pi = l)\left\{E_{Data} + \left(\left[\frac{K_u}{N}\right] - l\right)E_{idle}\right\}$$
$$E_{Data} = P_{idle}T_{IFS} + P_{Tx}T_{Data} + P_{Rx}T_{ACK}$$
$$E_{idle} = P_{idle}(T_{IFS} + T_{Data} + T_{ACK})$$

$$(21)$$

In CP, all nodes should perform pseudo random back-off delay after IFS whenever any data would be transmitted, but this would not affect the given time duration. Hence, we do not have to consider the random seed value for back-off delay when calculating the energy consumption described below.

$$E_{CP}^{N} = \lambda_e\left\{P_{Tx}T_{Data} + P_{Rx}T_{ACK} + P_{idle}\left(\frac{D_{CP}}{\lambda_e} - (T_{Data} + T_{ACK})\right)\right\}$$

$$(22)$$

A node generally keeps deep sleep mode in overall SP. However, when any emergency data is incurred, a node has to attach long preamble prior to the emergency data for successful reception from the PC which operates LPL mechanism in SP. Therefore, the energy consumption of a node in Sleep frame could be presented below.

$$E_{Sleep}^{N} = \lambda_e\left\{P_{Tx}(\varphi + T_{Data}) + P_{idle}\left(D_{sleep} - \varphi - T_{Data}\right)\right\} \\ + (1 - \lambda_e)P_{idle}D_{sleep} \\ = \lambda_e(P_{Tx} - P_{idle})(\varphi + T_{Data}) + P_{idle}D_{sleep}$$

$$(23)$$

Furthermore, we achieved simulation-based experiments to obtain realistic results compared with IEEE 802.15.6 protocol as shown in Figures 12 and 13 under the same scenario described in the previous subsection and Table 2. In this experiment, we operated a given network composed of one coordinator and several nodes during 100,000 s, and then we individually measured the total energy consumption of them.

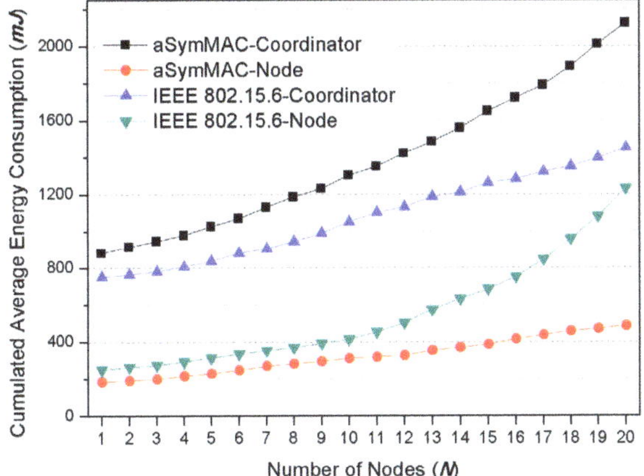

Figure 12. Cumulated energy consumption amount of both coordinator and node according to the increment of the number of nodes compared with IEEE 802.15.6 protocol.

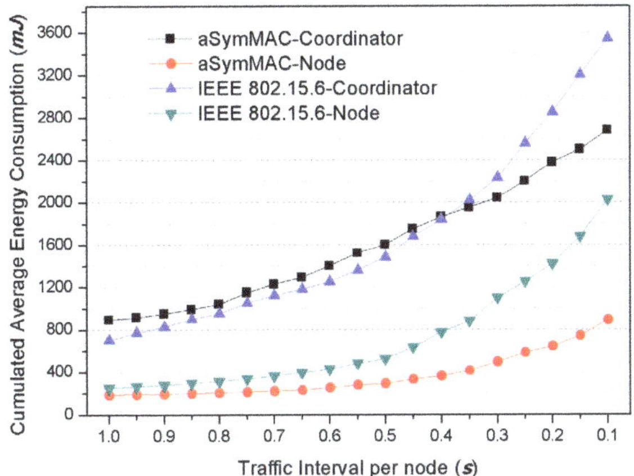

Figure 13. Cumulated energy consumption amount of both coordinator and node according to the traffic amount compared with IEEE 802.15.6 protocol.

Figures 12 and 13 show the cumulated energy consumption of the coordinator and the node employed in this simulation, according to the variable number of nodes and variable traffic amount. They also show the results of IEEE 802.15.6 protocol with the same conditions. In these figures, the proposed aSymMAC could not show the enhanced performance in PC compared with IEEE 802.15.6 protocol. However, it is not conclusively important in the real environment because a PC generally has the function of energy recharging and also has more powerful energy resource than the micro-sized on-body nodes.

Meanwhile, these figures present outstanding energy efficiency of the node side on the aSymMAC protocol compared with IEEE 802.15.6, and this is the key contribution of this research. In real environment, WBAN needs the extreme energy efficiency of the node side, but not PC. Figures 12 and 13 consequently prove that aSymMAC gives higher energy efficiency of the node than IEEE 802.15.6 and that the differentiation of energy efficiency between the coordinator and each node is not narrower on aSymMAC than IEEE 802.15.6. As a result, we can recognize that the main goal of this research has been accomplished.

From understanding these results from the overall figures in this section, we could find that the proposed MAC could achieve the enhancement of previous mechanisms even though some points might not be improved (e.g., the latency of normal data transmission and the energy efficiency of PC). However, in this case, the latency of normal data transmission and the energy efficiency of PC are not critically significant in the real environment.

6. Conclusions

Considering real environment of WBAN, the on-body node may have very constraint energy resource compared with the coordinator. In addition, emergency data incurred from a dangerous condition of the human body has to be preferentially delivered and reported to the coordinator or corresponding backend systems. These two elements have to be fundamentally required to meet the WBAN environment. We proposed a new scheme called aSymMAC protocol to satisfy the above two essential requirements. Moreover, the proposed scheme was evaluated with an analytic method and experiments focusing on the above two issues, and was compared with IEEE 802.15.6 which is the representative standard technology for WBAN environment. However, TDMA-based MAC protocols are vulnerable from the interference problem and aSymMAC could not be completely avoided from this problem. In advance, when many people who exploit aSymMAC-based healthcare system were spatially congregated, a solution to reassemble frame structure should be strongly required to address homogeneous interference from other WBAN. We will focus on this issue for further research.

Acknowledgments: This work was supported by the National Research Foundation of Korea (NRF) grant funded by the Korea government (MSIP) (No. 2016R1C1B1015863) and the Ministry of Education of the Republic of Korea and the National Research Foundation of Korea (NRF-2017R1D1A3B04034151).

Author Contributions: Jaeho Lee contributed by deriving the initial numerical and experimental results and by making the draft version of the paper. Seungku Kim was responsible for the main idea, theoretical analysis, coordination, and proofreading.

Conflicts of Interest: The authors declare no conflict of interest.

References

1. Patel, M.; Wang, J. Applications, Challenges, and Prospective in Emerging Body Area Networking Technologies. *IEEE Wirel. Commun.* **2010**, *17*, 80–88. [CrossRef]
2. IEEE 802.15.6: Wireless Body Area Networks. IEEE Standard for Local and Metropolitan Area Networks. 2012. Available online: http://ieeexplore.ieee.org/document/6161600 (accessed on 14 November 2017).
3. IEEE 802.15.4: Wireless Medium Access Control (MAC) and Physical Layer (PHY) Specifications for Low-Rate Wireless Personal Area Networks (WPANs). IEEE Standard Information Technology. 2006. Available online: http://ieeexplore.ieee.org/document/4299496 (accessed on 14 November 2017).
4. Li, H.; Tan, J. Heartbeat Driven Medium Access Control for Body Sensor Networks. *HealthNet* **2007**, *14*, 25–30.
5. Moulik, S.; Misra, S.; Chakraborty, C.; Obaidat, M.S. Prioritized payload tuning mechanism for wireless body area network-based healthcare systems. In Proceedings of the Global Commun (GLOBECOM), Austin, TX, USA, 8–12 December 2014; pp. 2393–2398.
6. Zhou, J.; Guo, A.; T Nguyen, H.; Su, S. Intelligent Management of Multiple Access Schemes in Wireless Body Area Network. *J. Netw.* **2015**, *10*, 108–116. [CrossRef]
7. LI, C.; LI, H.B.; Kohno, R. Reservation-Based Dynamic TDMA Protocol for Medical Body Area Networks. *IEICE Trans. Commun.* **2009**, *92*, 387–395. [CrossRef]
8. Fang, G.; Dutkiewicz, E. BodyMAC: Energy efficient TDMA-based MAC protocol for wireless, body area networks. In Proceedings of the 9th IEEE International Symposium on Communications and Information Technology ISCIT, Icheon, Korea, 28–30 September 2009; pp. 1455–1459.

9.	Khan, Z.A.; Sivakumar, S.; Phillips, W.; Aslam, N. A new patient monitoring framework and Energy-aware Peering Routing Protocol (EPR) for Body Area Network communication. *J. Ambient. Intell. Hum. Comput.* **2014**, *5*, 409–423. [CrossRef]

10.	Lee, J.; Kim, J.; Eom, D. A delay-tolerant virtual tunnel scheme for asynchronous MAC protocols in WSN. *Wirel. Pers. Commun.* **2013**, *70*, 657–675. [CrossRef]

11.	Buettner, M.; Yee, G.V.; Anderson, E.; Han, R. X-MAC: A Short Preamble MAC Protocol for Duty-Cycled Wireless Sensor Networks. In Proceedings of the 4th ACM International Conference on Embedded Networked Sensor Systems, Boulder, CO, USA, 31 October–3 November 2006; pp. 307–320.

12.	Polastre, J.; Hill, J.; Culler, D. Versatile Low Power Media Access for Wireless Sensor Networks. In Proceedings of the 2nd ACM International Conference on Embedded Networked Sensor Systems, Baltimore, MD, USA, 3–5 November 2004; pp. 95–107.

13.	TI CC1100 Datasheet. Available online: http://www.ti.com (accessed on 14 November 2017).

 *applied sciences*

Article

Virtual Reality-Wireless Local Area Network: Wireless Connection-Oriented Virtual Reality Architecture for Next-Generation Virtual Reality Devices

Jinsoo Ahn [1], Young Yong Kim [1] and Ronny Yongho Kim [2,*]

[1] School of Electrical & Electronics Engineering, Yonsei University, Seoul 03722, Korea;
    gumgoki@yonsei.ac.kr (J.A.); y2k@yonsei.ac.kr (Y.Y.K.)
[2] Department of Railroad Electrical and Electronic Engineering, Korea National University of Transportation,
    Gyeonggi 16106, Korea
*   Correspondence: ronnykim@ut.ac.kr; Tel.: +82-31-460-0573

Received: 15 November 2017; Accepted: 27 December 2017; Published: 3 January 2018

Featured Application: Virtual reality headsets and controllers with wireless connection.

Abstract: In order to enhance the user experience of virtual reality (VR) devices, multi-user VR environments and wireless connections should be considered for next-generation VR devices. Wireless local area network (WLAN)-based wireless communication devices are popular consumer devices with high throughput and low cost using unlicensed bands. However, the use of WLANs may cause delays in packet transmission, owing to their distributed nature while accessing the channel. In this paper, we carefully examine the feasibility of wireless VR over WLANs, and we propose an efficient wireless multiuser VR communication architecture, as well as a communication scheme for VR. Because the proposed architecture in this paper utilizes multiple WLAN standards, based on the characteristics of each set of VR traffic, the proposed scheme enables the efficient delivery of massive uplink data generated by multiple VR devices, and provides an adequate video frame rate and control frame rate for high-quality VR services. We perform extensive simulations to corroborate the outstanding performance of the proposed scheme.

Keywords: multiuser; OFDMA; Virtual Reality; wireless LAN; wireless VR

---

1. Introduction

Network operators and system administrators are interested in the mixture of traffic carried in their networks for several reasons. Knowledge about traffic composition is valuable for network planning, accounting, security, and traffic control. Traffic control includes packet scheduling and intelligent buffer management, to provide the quality of service (QoS) needed by applications. It is necessary to determine to which applications packets belong, but traditional protocol layering principles restrict the network to processing only the IP packet header.

Virtual reality (VR) devices are novel, and attractive consumer electronics that can provide an immersive VR user experience (UX) [1,2]. In order to enhance the UX of VR services, there have been significant efforts to enhance not only its video and audio quality and interaction delay, but also the convenience of VR device connections to VR consoles, or VR-capable personal computers (PCs). Therefore, consumers in the VR market require high-resolution and comfortable VR devices. In order to provide a comfortable VR service environment without the need for wires, a wireless communication scheme with low latency needs to be employed for VR devices. However, high-resolution features are not only a challenge with respect to imaging equipment, but also for wireless interfaces. Therefore, we need to find the trade-off between these two features, i.e., reliable wireless connection and high-resolution video.

Wireless local area network (WLAN) is the most popular unlicensed-band wireless communication interface, which has a low cost while achieving high data throughput. Because some categories of IEEE 802.11 are designed to replace wired video interfaces, including high-definition multimedia interface (HDMI), IEEE 802.11-based WLAN can provide very high data rates.

In order to meet the high data-rate requirement of high-resolution video transmission, the IEEE 802.11 working group extended IEEE 802.11 standards to support the 60-GHz frequency band with a wide bandwidth. IEEE 802.11ad is the amendment standard, to operate IEEE 802.11 in the 60-GHz frequency band. In particular, IEEE 802.11ad utilizes the 2.16-GHz bandwidth to achieve a high data rate. However, IEEE 802.11ad has a relatively short communication range, owing to high-frequency band operation under indoor environments [3,4]. IEEE 802.11ay is the enhanced version of IEEE 802.11ad, with the support of channel bonding and multiple spatial streams [5].

Although these 60-GHz WLAN standards can be considered as a wireless VR interface for high-resolution video transmission, future VR systems will require real-time interactive control between multiple VR users. Some applications related to gaming industries have been adopting multiuser augmented reality (AR) systems to provide enhanced gaming experiences in the living room [6]. Because VR consumers have already experienced these multi-user AR systems, multi-user VR also needs to be provided to satisfy the needs of VR consumers. The sharing of VR experiences with nearby users is expected to provide much more immersive VR UX to VR consumers [6–8].

Wireless multi-user VR systems based on IEEE 802.11 standards can be described as in Figure 1, which shows the elements of wireless multi-user VR systems, VR data flows, and delay components of VR services. In order to provide an immersive VR interaction experience, each component of the VR system shall provide proper feedback, based on its sensing data. From delay components of Figure 1, the VR interaction delay of a wireless VR system, T_{vr}, can be described as follows.

$$T_{vr} = T_{sensing} + T_{proc1\_in\_device} + T_{transfer\_UL} + T_{proc\_in\_PC} + T_{transfer\_DL} + T_{proc2\_in\_device} \tag{1}$$

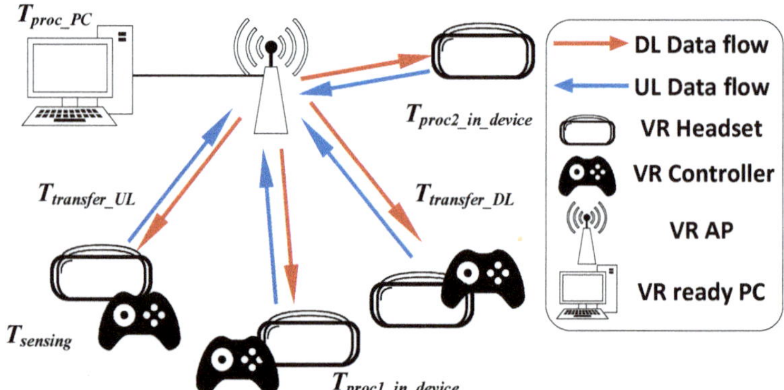

Figure 1. Wireless multiuser virtual reality (VR) system, based on IEEE 802.11 system.

VR devices shall track motion and command of users by using sensing components. The sensing components may cause a delay, depending on their sensing performance. This delay is considered as $T_{sensing}$. $T_{proc1\_in\_device}$ is a processing delay for a processor unit in VR devices, which handles sensing data and generates data packets. The generated data packets are transmitted to the associated VR computing device over a WLAN link in the proposed system. One-hop wireless packet delivery over a WLAN link causes a delay, $T_{transfer\_UL}$, which could be a relatively large value depending on the wireless channel condition, including channel congestion caused by channel contention. $T_{transfer\_UL}$ is the most dominant delay component in the proposed multi-user VR system. The VR computing device, which is generally a PC, requires a processing delay, $T_{proc\_in\_PC}$. In order to provide seamless VR UX by minimizing a processing delay, high-processing performance is preferred. The VR computing device generates VR feedback packets, including VR video data, and transmits them to its associated VR devices over a WLAN link. The delay caused by this WLAN transmission is considered as $T_{transfer\_DL}$. Since the transmission causing the delay of $T_{transfer\_DL}$ is from one node (VR computing device) to multiple nodes (wearable VR devices), and the transmission causing the delay of $T_{transfer\_UL}$ is from multiple nodes (wearable VR devices) to one node (VR computing device). $T_{transfer\_DL}$ can be more easily controlled than $T_{transfer\_UL}$. The downlink delay, $T_{transfer\_DL}$, is the second most important delay component in the proposed system. The VR devices decode the packet and operate to generate sensory feedback, which causes a processing delay, $T_{proc2\_in\_device}$. For instance, a VR headset decodes VR video image and perform video enhancement procedure, for a seamless and immersive UX. $T_{proc2\_in\_device}$ is a delay component for this kind of hardware processing, to provide sensory feedback.

IEEE 802.11 systems are designed as contention-based, channel access, wireless communication systems [9]. Because of the properties of contention-based channel access, the performance of IEEE 802.11ad/ay systems degrades dramatically as the number of wireless stations (STAs) increases [10]. In other words, even though most advanced WLAN protocols and VIDEO codecs are utilized, supporting multi-user VR with low latency is almost impossible, owing to the multiple access inefficiency of WLAN.

VR devices need to upload their sensing information to trace user position and pose frequently, and each VR device usually keeps track of its position with a 1000-Hz sensing rate. This means that in multi-user VR, small uplink frames are generated very frequently, by multiple VR devices. Severe WLAN channel contention is caused by an overwhelming number of small VR control frames, leading to a very long channel access delay, which makes the operation of multi-user VR over WLAN impossible. Such problems cannot be solved by conventional IEEE 802.11 distributed coordination function (DCF) and enhanced distributed channel access (EDCA), which do not guarantee frame delivery delay [9].

In order to provide multi-user VR services over WLANs, the channel access delay needs to be minimized, and the frame rate of VR video and the arrival rate of uplink (UL) frames should be adaptively adjusted, depending on the wireless environment. Since both the technical progress of frame-interpolation schemes [11–16] and the frame-interpolation module are expected to be commonly employed in next-generation VR headsets [17], users could have stable high-refresh-rate vision with a lower downlink (DL) VR video frame rate. Sensing data, which is not fed back to computing machines, including VR-ready PCs, can be utilized by the frame-interpolation module to generate interpolation frames, with accurate moving data from the user. These interpolation frames are not based on future frames [11], because these are real-time video frames. The interpolation frames are generated from past frames and motion track data. Moving its latest frame to opposite vector of sensor data is the easiest method of generating interpolation frames.

In order to assist the interpolation frame generation, some network characteristics, e.g., the video frame arrival rate and control frame delivery rate, need to be provided to VR devices. When there is a mismatch between the visual and vestibular systems, VR sickness can result. This means that if the VR vision in VR displays cannot reflect the real movement of users, the user experience may be degraded. In order to prevent VR sickness, accurate VR frame interpolation operations are required. Therefore, in the network-condition-based VR frame delivery proposed in this paper, the video frame interpolation procedures are very important, in order to reduce VR sickness. The relationship between the received video frames and interpolated video frames is shown in Figure 2. In this paper, interpolated video frames do not refer to the video frames generated by the graphic processor of a VR PC or a VR console. Here, only frames that are generated by VR headsets after receiving video frames from a VR access point (AP) are referred to as interpolated video frames, which can be generated using received frames and motion track information. As shown in Figure 2, the processing unit in a VR headset shall move the last video frame to the reverse direction of user's motion vector, to generate interpolated video frames. The interpolated video frames provide immediate visual responses that solve the mismatch between visual and vestibular systems. In wireless multi-user VR systems, since wireless links with inefficient multi-user channel access performance are bottlenecks, which cannot provide a sufficiently high data rate for a high video frame rate, many interpolated frames are generated. Because of such problems, next-generation wireless multi-user VR systems should be designed to be tightly coupled with wireless systems. The next-generation, wireless, multi-user VR systems should optimize their VR video image and motion tracking rate, considering the wireless link status. Based on the above observation, in order to design a high-quality multi-user VR system over WLANs, both wireless link optimization, which enhances the wireless channel access efficiency, and tight-coupled VR optimization with a wireless system, which prevents unnecessary resource wastage, should be considered. In this study, to provide high-quality VR UX in a multi-user WLAN VR service, we consider both the multi-user wireless link efficiency enhancement and VR optimization tightly coupled with a wireless system.

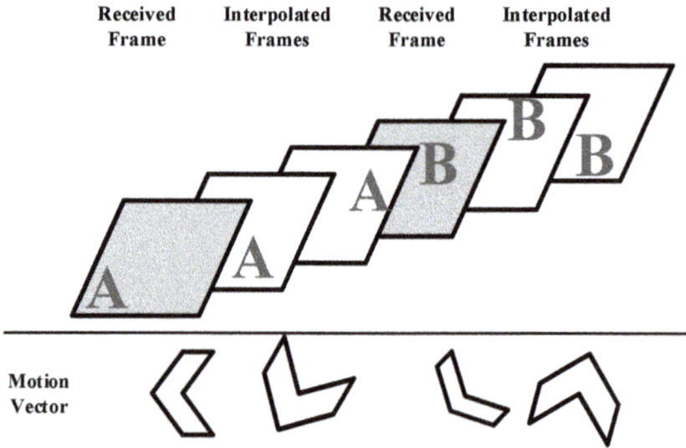

Figure 2. Motion sensing based video interpolation in VR headsets.

This paper is an extension of one that proposed delay-oriented VR mode that could be utilized by a VR AP [18]. The delay-oriented VR mode is included in this paper as a trigger-based channel access method. This paper proposes a novel wireless multi-user VR protocol structure, as well as specific channel access and system control schemes, to support multi-user VR systems over WLAN, including the delay-oriented VR mode. In addition to the novel structure and the enhanced channel access and control schemes, in this paper, we propose connection-recovery algorithms for a seamless VR UX.

The rest of this paper is organized as follows. In Section 2, we explain the proposed system architecture and protocol design, including the connection–recovery algorithm. In Section 3, we investigate the system performance of the proposed VR architecture and multi-user VR schemes, by performing extensive simulations. We also examine the delay and packet loss rate (PLR) performances in various simulation scenarios. Finally, Section 4 explains the reason why conventional EDCA, which is utilized in WLAN systems, cannot handle wireless multi-user VR applications, but the proposed system can handle them.

2. Architecture and Protocol

2.1. IEEE 802.11-Based Wireless VR System Architecture

The proposed multi-user VR systems with wireless interfaces consist of multiple IEEE 802.11 medium access control (MAC) layers and physical (PHY) layers. In order to accommodate multiple

IEEE 802.11 protocols, we now propose a novel VR convergence layer (VRCL) and its interworking scheme with a station management entity (SME).

The network architecture for multi-user VR systems needs to meet the requirements of very high throughput and low latency. In order to satisfy the high data throughput requirement of high-resolution video images, we can utilize 60-GHz standards, i.e., IEEE 802.11ad/ay. These amendment standards are designed for wireless high-resolution image devices. Because immersive VR UX could be achieved by these high-resolution video images, the use of 60-GHz standards is inevitable in multi-user VR scenarios. However, although these high-throughput wireless standards are utilized, with a combination of UL and DL transmission in multi-user scenarios, the effective throughput and delay performance would deteriorate. This means that intra-basic service set (BSS) channel contention should be controlled by wireless VR protocols. Without such control algorithms and additional channel resources, VR experience cannot be guaranteed in wireless network environments.

If there are only DL video frames transmitted by a VR AP, network degradation may not occur. Multiple VR devices connected to an AP should transmit their motion tracking data and control data very frequently, i.e., 1000 Hz per device. Those uplink motion tracking and control data are problematic, because frequent motion tracking and control data can cause large channel access delays and throughput degradation. In order to resolve such an uplink data contention problem, multiuser control channels conforming to the wireless standard should be utilized for multi-user VR networks. IEEE 802.11ax is the most representative standard that supports multi-user network in the 5-GHz frequency band [19]. This means that because the 5 GHz channels of the IEEE 802.11ax standard do not interfere with 60-GHz frequency channels, a VR AP can accommodate multi-user uplink traffic very efficiently.

As a result, VR devices including VR APs need to have special multi-standard protocol architecture, described in Figure 3, to support VR connections based on WLANs. IEEE 802.11ax is the amendment standard for highly efficient WLAN in multiple-device scenarios. IEEE 802.11ax defines a trigger frame to accommodate multiple uplink frames from multiple devices simultaneously [19]. In many scenarios, it may be utilized in parallel with a conventional single-frame transmission. If the trigger frame requests stations to transmit its UL data, each station transmits its data without additional channel access delays. This trigger frame is able to substantially reduce the contention delay of WLAN systems.

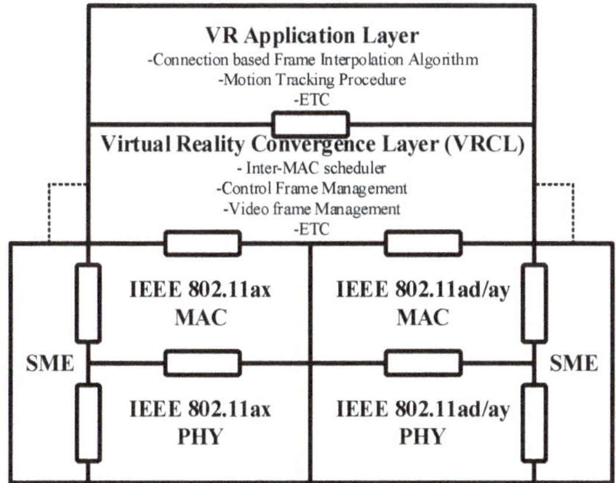

Figure 3. Proposed protocol architecture of a wireless, multi-user VR system.

The IEEE 802.11ad amendment standard is designed to utilize the 60-GHz frequency band, which provides wide bandwidth and high throughput. IEEE 802.11ay is an enhanced version of IEEE 802.11ad, and provides four times the bandwidth using channel bonding and additional spatial streams [5]. Because of the wide bandwidth, the 60-GHz standard is able to provide very high data rates over short distances. Therefore, in order to adequately utilize the high data rate to support multi-user VR, the channel inefficiency caused by channel contention should be minimized by separating the UL transmission and DL transmission.

The VR application layer described in Figure 3 is a protocol layer that provides VR images and control information on VR devices. VR video frames are generated using the frame rate that is reported by VRCL, and a detailed explanation on VRCL is provided in Section 2.2. The generation rate of the VR video frame is restricted by the VRCL, and it prevents the VR application layer from generating meaningless video frames. For VR controllers and VR headsets, motion-tracking information measured by sensors in VR devices is accommodated and utilized in this VR application layer. VR devices, especially VR headsets, would generate interpolation frames in this layer. The interpolation frame is the frame that needs to be displayed between the received real video frames delivered by a VR AP. These interpolated frames could be generated by performing many effective interpolation algorithms [11–16]. In this paper, VR video interpolation frames need to be based on motion-tracking information that is measured by VR sensors in VR devices [20–22]. Because this situation was not previously defined, further optimized interpolation methods should be studied.

The convergence layer described in Figure 3 is a protocol layer that enables multiple network standard convergence, as well as some special information for the VR application layer. VR videos that are generated by a VR PC are delivered to a VR AP, and the convergence layer in the VR AP determines the transmission interval of video frames based on network conditions, and reports the rate to the VR application layer of the VR PC. If VR video frames are transmitted without these considerations for network conditions, users would suffer poor UX, owing to the large delay time. Similar to the DL VR video frame transmission, the convergence layer also controls the UL frame delivery rate, based on its network condition. This would reduce network congestion or the required network performance of VR devices. As a result, the convergence layer prevents these catastrophic situations by controlling the frame delivery rate.

Although some motion-tracking information cannot be delivered, depending on the decision of the convergence layer, the convergence layer still provides motion-tracking information to the VR application layer in the VR headset, to generate interpolation frames. These interpolation frames should be generated considering motion-tracking information, in order to prevent VR sickness. The number of interpolation frames that need to be generated before the next frame may be predicted by the frame arrival rate information obtained from the convergence layer.

The station management entity (SME) is used for the accommodation and delivery of parameters for each network layer [9]. In some cases, frame off-loading could be performed by controlling the frame-delivery interval based on PHY and MAC layer parameters. The packet loss rate information and frame interval information are key sets of information delivered by the SME.

Each MAC and PHY layer follows its own standards. The convergence layer controls and schedules all frames into those multiple MAC layers properly. For example, DL data frames can be scheduled in the IEEE 802.11ad/ay MAC and UL data frames can be scheduled in the IEEE 802.11ax MAC. Because IEEE 802.11ax is a 5-GHz standard with multi-user support, it is a suitable protocol for the UL transmission protocol in multi-user VR systems. Figure 4 shows how IEEE 802.11ax could accommodate multiple frames, using orthogonal frequency-division multiple access (OFDMA). The trigger frame is transmitted by an AP, to instruct stations that have UL frames when and where to transmit their UL frames. In usual IEEE 802.11ax scenarios, the trigger frame would contend with other UL frames to guarantee the opportunity for all stations to access the channel. However, a VR AP requires a very tight delay property, and its traffic pattern is very regular—it is a special-purpose AP for VR devices. This means that for associated devices, there is no need for channel contention to guarantee opportunities for channel access. In other words, in order to fully utilize the UL OFDMA of IEEE 802.11ax for multi-user VR services, single-user UL transmission should be regulated. Single-user

UL transmission can be regulated by the multi-user EDCA procedure in 802.11ax [19]. By using a new set of EDCA parameters, which is used by STAs in a multi-user BSS, AP can set STAs to have very low-priority EDCA parameters. Such low-priority EDCA parameters make STAs' access time for a single-user UL transmission long, and the AP can transmit a trigger frame for UL OFDMA procedure while the STAs are waiting for the single-user UL transmission.

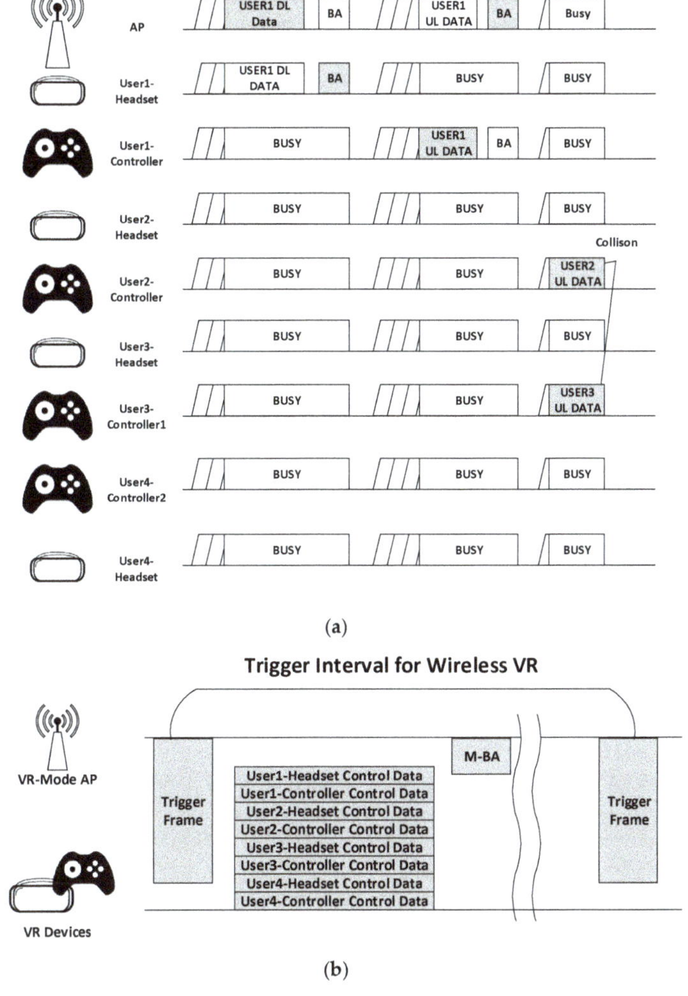

Figure 4. Distributed coordination function (DCF) and the proposed delay-oriented VR mode operation for wireless VR system. (**a**) Wireless multiuser VR with conventional wireless local area network (WLAN); (**b**) Wireless multiuser VR with the proposed delay oriented VR mode WLAN.

2.2. Protocol Design

The VRCL should encapsulate interpolation and frame rate information, using a VR video frame in an aggregated MAC protocol data unit (A-MPDU). The VRCL would control the UL and DL frame arrival rate for WLAN systems based on the frame rate of the original VR video rate and the wireless environment. Owing to VRCL, a VR AP and VR devices could utilize IEEE 802.11 family standards that provide MAC and PHY layers without standard modification. VRCL is the only additional layer for a multi-user VR network interface. VR applications do not require any additional network control features, and are required only to generate proper VR video frames, based on information delivered by the VRCL. The VRCL provides its UL control information rates and the required DL video frame to the VR application layer. Because the VRCL would discard VR frames that could cause large network delays on VR devices, the VR application does not need to waste its resources on unnecessary video frames. For this reason, the VRCL and VR AP need to perform the function of a VR system controller.

Not only does the VRCL provide information that can be used to control VR video frames to the VR application layer, but VRCL also controls its network operations based on information that is provided from the SME. The packet loss rate, received signal strength indication (RSSI), and modulation and coding scheme (MCS) index information are the representative information observed by VRCL. Based on the observed information, VRCL could control the MCS level, channel bandwidth, frame arrival rate, number of users supported by the VR AP, and so on. The 60-GHz and 5-GHz standards may be utilized by VRCL, because of its efficient multi-user VR operation. The main purpose of the 60-GHz wireless link, in this paper, is the delivery of a high-resolution VR video frame. Because that delivery requires a large bandwidth, the use of the 60-GHz standard is inevitable. A 5-GHz standard is employed to accommodate the multi-user UL frame, by utilizing the multi-user UL OFDMA procedure.

Although the required data rate of multi-user uplink sensing data is not very high, the frame arrival rate of the sensing data frames is relatively high. Usually, the video frame rate ranges from 90 Hz to 120 Hz, but the motion sensing rate is 1000 Hz in current-generation, wired VR devices. Because of this, a small-size UL data frame could spoil the overall wireless VR system, in spite of its small required data rate. This causes a large channel access delay on the DL VR video frame, unless the DL and UL are separated. By separating the UL sensing data transmission from the 60-GHz standard, DL VR video frames do not contend with other frames from VR devices. If a VR video frame requires additional transmission time, owing to its poor channel condition or the increasing number of VR users, the VRCL can control the video frame rate based on its packet loss rate. The VRCL in a VR AP encapsulates the video frame rate information that is used to generate VR video frames. The VRCL in VR devices extracts the information from the received frames and delivers the information to the VR application, in order to generate a video interpolation frame based on its received video frames and motion-sensing information.

Unless the channel has high coexistence issues, the DL frame may not have large channel access interference. However, in the WLAN multi-user VR scenario, UL frames always suffer large contention delays, owing to frequent channel access with the contention-based channel access method. In order to solve the problem, a VR AP should support the UL frame accommodation by transmitting a trigger frame, which is defined in the IEEE 802.11ax standard. In order to maximize the UL, the multi-user frame accommodation channel access of each VR device should be prohibited. The channel access may be prohibited by setting EDCA parameters, including AIFSN and CW, to very large values, and maximum values are particularly recommended. Setting the EDCA parameters does not require any modification to IEEE 802.11 standards, and manufacturers could configure the EDCA parameters easily.

2.3. Algorithm

The VRCL in a VR AP could perform DL VR video frame rate control and UL VR sensing frame rate control, depending on its wireless connection status. Figure 5 shows how the VRCL controls its DL VR video frame rate. The manufacturer of a VR system could set its packet loss rate thresholds and corresponding VR video frame rate. A video refresh rate of 90 Hz is usually used in current-generation VR systems, but next-generation VR systems may support a refresh rate of at least 120 Hz. This means that "refresh_rate_1" in Figure 5 needs to be set to its native refresh rate of the VR display. Refresh rate parameters with larger index numbers should be set to a larger value than refresh rate parameters with smaller index numbers. For "DL_PLR_threshold" parameters, as listed in Figure 5, the same rules should be applied, and a parameter with a larger index needs to be set to a larger value. The proposed algorithm in Figure 5 aims to solve the connection problem of wireless multi-user VR systems, by controlling the required channel throughput. If the channel condition is worse than "DL_PLR_threshold_n" (see Figure 5), the VR AP should perform a recovery procedure. The recovery procedure can be modified by the manufacturer, but the use of alternating channels and 5-GHz off-loading is at least recommended. Figure 6 shows an algorithm that controls the trigger frame transmission rate. A high value for "Trigger_rate_VR" (see Figure 6) indicates a small trigger frame interval, based on the PLR history of the UL sensing data frame. Because associated VR devices never perform EDCA procedures, they are disabled by the VR AP, and only overlapped BSS (OBSS) stations could cause channel collisions and channel interference. Similar to DL VR video frame rate control, the VRCL controls the frame rate and performs recovery procedure in poor channel conditions. In this paper, channel alternation and bandwidth modifications are recommended for the recovery procedures.

The VRCL in a VR headset could receive the DL VR video frame rate and UL VR sensing frame rate information from the VR AP. Based on each set of information, the VR application could generate an interpolation frame that is utilized during the DL VR video frame interval. Each interpolation frame should utilize motion-sensing information, even though the information was not delivered to the VR AP. From the motion-sensing information, the VR application layer should generate a motion vector, and the reverse vector of the generated motion vector should be applied to the interpolation frame.

```
while
    for(i = 1 to number_of_user)
        if PLR_DL < DL_PLR_threshold_1
            Frame_rate_VR = Refresh_rate_1
            Frame_MCS_VR = MCS_1
        elseif PLR_DL < DL_PLR_threshold_2
            Frame_rate_VR = Refresh_rate_1
            Frame_MCS_VR = MCS_2
                ⋮
        elseif PLR_DL < DL_PLR_threshold_n
            Frame_rate_VR = Refresh_rate_n
            Frame_MCS_VR = MCS_n
        else
            Vendor_specific_recovery procedure{
                if another 60GHz band channel is available
                    Channel alternating procedure (channel information)
                elseif 5GHz offloading is available
                    Frame offloading scheduler (Refresh_rate_n)
                else
                    Disconnect weakest connection user (user_profile)
                    Wireless VR connection failure message (user_profile)
                    Modify user profile (user_profile)
                    Number_of_user--
                endif
            }
        endif
    end
end
```

Figure 5. Proposed algorithm to adjust downlink (DL) VR video frame rate, and examine its feasibility of VR capacity for a given user profile.

```
while
    for(i = 1 to number_of_user)
        if PLR_UL < UL_PLR_threshold_1
            Trigger_rate_VR = Sensing_rate_1
            Channel bandwidth = maximum
        elseif PLR_UL < UL_PLR_threshold_2
            Trigger_rate_VR = Sensing_rate_2
            Channel bandwidth = maximum
                ⋮
        elseif PLR_UL < UL_PLR_threshold_n
            Frame_rate_VR = Refresh_rate_n
            Channel bandwidth = maximum
        else
            Vendor_specific_recovery procedure{
                if other 5GHz band channel is available
                    Channel alternating procedure (channel information)
                elseif (Channel RSSI < RSSI_threshold) && (Channel rate > required
                throughput) && (Channel bandwidth > minimum)
                    Channel bandwidth = Channel bandwidth / 2
                else
                    Disconnect weakest connection user (user_profile)
                    Wireless VR connection failure message (user_profile)
                    Modify user profile (user_profile)
                    Number_of_user--
                endif
            }
        endif
    end
end
```

Figure 6. Proposed algorithm to adjust uplink (UL) VR sensing information report rate, and examine its feasibility of VR capacity for a given user profile.

3. Results and Analysis

3.1. Performance Metric Definition

In order to support a 1000-Hz sensing rate, we set the basic UL sensing frame rate to 1000 Hz. This means that VR devices generate their sensing frame every 1 ms. In this paper, we applied a separate WLAN structure that we proposed. In addition to the structure proposal, we compared the delay and packet loss rate properties of our proposed algorithm in various situations with conventional EDCA cases.

The main focus of this analysis is $T_{transfer\_UL}$ of Equation (1), because $T_{transfer\_UL}$ is relatively large despite its small UL frame size. $T_{transfer\_UL}$ can be further decomposed as follows:

$$T_{transfer\_UL} = T_{preamble} + T_{data} + T_{contention} + T_{IFSs} + T_{sync} \tag{2}$$

where $T_{contention}$ includes the back-off time, as well as channel-busy duration during a back-off procedure, when the delays are caused by channel access of initial transmission and retransmissions. Because T_{IFSs} is a fixed-time duration defined in the IEEE 802.11 standard, it cannot be changed. $T_{preamble}$ is also determined by the IEEE 802.11 standard. However, by using UL OFDMA transmission, only one $T_{preamble}$ is required for multiple stations, because UL stations transmit their preamble simultaneously at the beginning of the UL transmission. T_{data} is the time duration required for over-the-air data transmission. In the EDCA case, the cumulative $T_{preamble}$ would require a much longer duration, even though it does not include any information for VR applications. Therefore, the proposed scheme utilizes the multi-user physical frame structure to reduce the effective $T_{preamble}$ although it requires a longer T_{data}, owing to the small size of the resource unit (RU) for a single device. The RU is a unit of the frequency resource for OFDMA transmission, and the size and location of the RU for UL OFDMA transmission can be indicated by the trigger frame.

The proposed wireless multi-user VR system can reduce not only the effective $T_{preamble}$, but also $T_{contention}$. Because not all VR devices in the VR AP BSS are allowed to transmit their UL control data without receiving the trigger frame, there would be no contention. On the other hand, because VR devices cannot transmit their UL data without a trigger frame from the VR AP, we added a new delay factor T_{sync}. T_{sync} is a delay parameter caused by the time gap between the trigger frame reception and VR traffic generation.

In this paper, we assumed the existence of nine wireless VR devices, including VR controllers and VR headsets. Additional specific simulation parameters are shown in Table 1. Because of the effect of preamble overhead reduction, a duration that is nine times that of $T_{SU\text{-}UL}$ could be replaced by a single $T_{MU\text{-}UL}$ duration in the proposed multi-user VR mode case. T_{AIFS} is the required time duration for channel sensing before trying to perform the channel access procedure. M_{Queue} is the size of the queue parameter for each device. If the M_{Queue} is filled, and an additional frame comes to the queue, the IEEE 802.11 system would discard the last frame, which is calculated as a packet loss that increases the packet loss rate. $\lambda_{Trigger}$ is the trigger frame delivery rate controlled by VRCL, as mentioned above. VRCL could configure this value to provide seamless VR UX.

Table 1. Simulation Parameters.

Symbol	Description	Value
N_{VR}	Number of VR consumer devices	9 (3 headsets, 6 controllers)
$T_{UL\text{-}SU}$	Transmission opportunity time of a control frame in single-user transmission case	90 μs
$T_{UL\text{-}MU}$	Overall transmission opportunity time for UL multiuser transmission case	540 μs
T_{AIFS}	Channel sensing duration before initiating channel access procedure	36 μs
M_{Queue}	Size of frame queue in number of frames	1000 frames
$\lambda_{Trigger}$	Trigger Frame delivery rate (Configurable)	100 Hz to 1000 Hz, 1000 Hz is default
λ_{Motion}	Motion sensing rate	1000 Hz
N_{retx}	Maximum number of retransmission (Configurable)	0 or 2
N_{OBSS}	Number of interference devices (Configurable)	0 or 9
λ_{OBSS}	Frame arrival rate of each interference devices.	1 Hz
T_{OBSS}	Transmission opportunity time of an interference frame	10 ms = 10,000 μs

Even though λ_{Motion}, which is the motion sensing rate, is fixed to 1000 Hz by hardware, the VRCL controls its delivery rate, and undelivered information is used to generate interpolation frames in a VR headset. N_{retx} is a parameter for the maximum number of retransmissions (retx). Because VR systems are real-time applications, a large number of retransmissions is not suitable. In this study, we simulated cases involving no retransmissions up to two retransmissions. In order to test the robustness of the proposed system, we considered the presence of WLAN interference devices. N_{OBSS} is a parameter that shows the number of WLAN interference devices that exist. In the no-interference cases, N_{OBSS} is set to 0; otherwise, nine interferences devices are assumed. In order to measure the actual effect caused by interference devices, the reasonable rate of WLAN frames should be set to WLAN interference devices. In this paper, λ_{OBSS}, which refers to the frame arrival rate of each WLAN interference device, is set to 1 Hz. This means that every 1 s, WLAN interference devices generate a frame to transmit. The frame generated by WLAN interference devices has an air-time duration of T_{OBSS}, and in this study, we set it to 10 ms.

3.2. Simulation Results

In conventional EDCA cases, nine devices generated their UL frames containing motion-sensing information every 1 ms. For conventional EDCA cases, the packet loss rate and $T_{transfer\_UL}$ are shown in Figure 7. Because VR systems require real-time processing, the number of maximum retransmissions did not need to be set to a large number. If it was set to 2, as shown in Figure 7a, it would lower the PLR for a short time, but then increase to 0.4 PLR shortly afterwards. From the start, the PLR observed in conventional EDCA cases was not usable for consumer VR devices. In the no-retransmission case, as shown in Figure 7b, even though there was a large PLR and relatively small delay, not only was the PLR destructive, but the delay also could not be used. After 8 s, the delay value increased over 1 s for the motion-sensing data delivery. From the results of Figure 7, we can observe that conventional EDCA procedures cannot afford wireless multi-user VR systems. At a minimum, wireless VR systems require ms-level delays and a very low PLR for reasonable user experience.

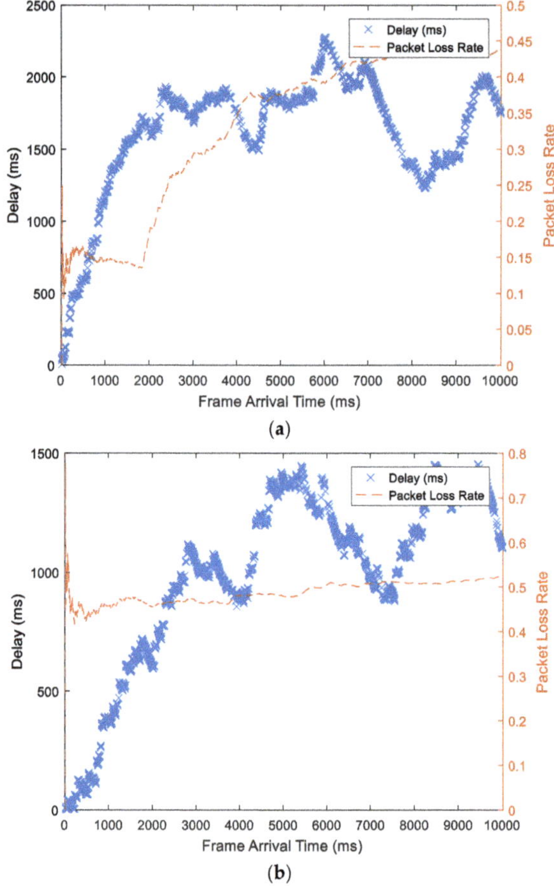

Figure 7. Delay and packet loss rate (PLR) for proposed WLAN structure with conventional enhanced distributed channel access (EDCA). (**a**) 9 VR devices with max number of retx = 2; (**b**) 9 VR devices without retx.

The proposed VR AP disables the conventional EDCA procedure, which can cause tragic delays and PLR results as shown in Figure 7. If there are no interference devices, which are usually called

OBSS devices, VR AP is the only wireless device that initiates the channel access procedure. In this case, a VR AP always obtained transmission opportunity without channel contention and back-off procedure, because the channel was idle during every channel access time period. This meant that the VR AP did not require an additional contention delay that is caused by the back-off procedure.

Although the proposed architecture and schemes work very well, as shown in Figure 8, the channel condition could be worse, owing to interference devices in the 5-GHz frequency band. As opposed to 60 GHz band radio, the 5-GHz band radio has a relatively long transmission range, interference range in this case, and 5-GHz frequency band devices are widely used by ordinary consumers. This means that a VR AP should consider the influence of interference. In order to consider and measure the effect of channel interference, in this paper we considered nine additional IEEE 802.11 devices that utilized the 5-GHz band. Each device generated its data frame with an air time of 10 ms every 1 s. The result is shown in Figure 9, and as opposed to the no-interference case, these results showed relatively large delays. Most frame-delivery delays were under 10 ms, but a large delay jitter was observed. If we set the maximum number of retransmission parameters to 0, the delay jitter would be decreased, as shown in Figure 9b.

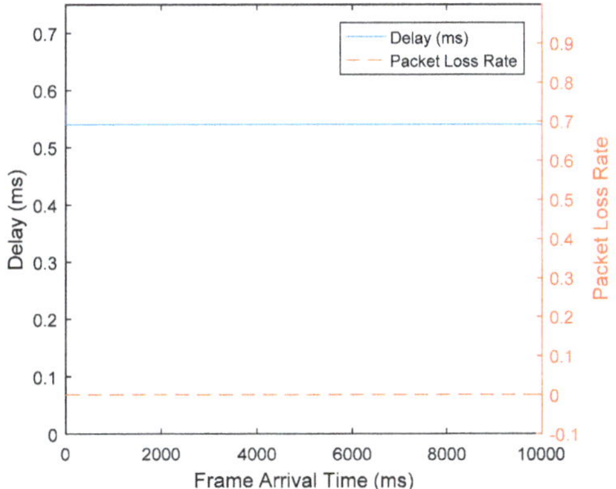

Figure 8. Delay and PLR for proposed WLAN structure with proposed schemes in a no-interference environment.

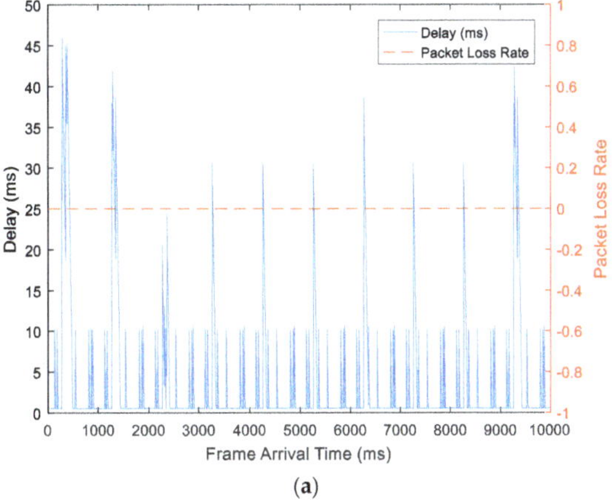

(a)

Figure 9. *Cont.*

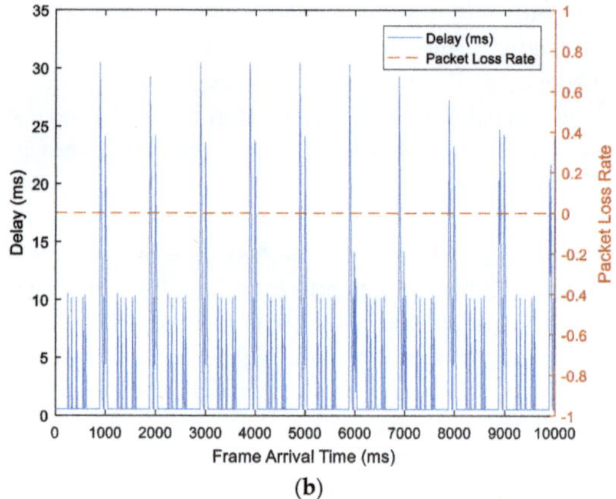

(b)

Figure 9. Delay and PLR for proposed WLAN structure with proposed schemes in a severe-interference environment. (**a**) Nine interference devices with max # of retx = 2; (**b**) Nine interference devices without retx.

Even though the 10-ms delay performance was relatively good for multi-user VR, in the severe-interference case, the delay jitter still needed to be minimized. The proposed algorithm to control the trigger frame delivery rate in Figure 6 could handle this severe interference case. Because the proposed VR system could realize variable-motion sensing information delivery rates, owing to its VR video interpolation module in VR headsets, its sensing data report rate could be adjusted to a small value. In this study, we assume a 100 Hz to 1000 Hz sensing data report rate. If the channel conditions are very poor, the VR AP should set its trigger frame delivery rate to 100 Hz. In order to measure the effect of controlling the trigger frame, we simulated the case with a 100-Hz trigger frame delivery rate in the severe-interference scenario. The result is shown in Figure 10, and it showed that in the no-retransmission case, all of the frames were delivered in a 10-ms delay. This means that the proposed wireless multi-user VR system guarantees a 10-ms wireless uplink delay, even in the presence of severe channel interference.

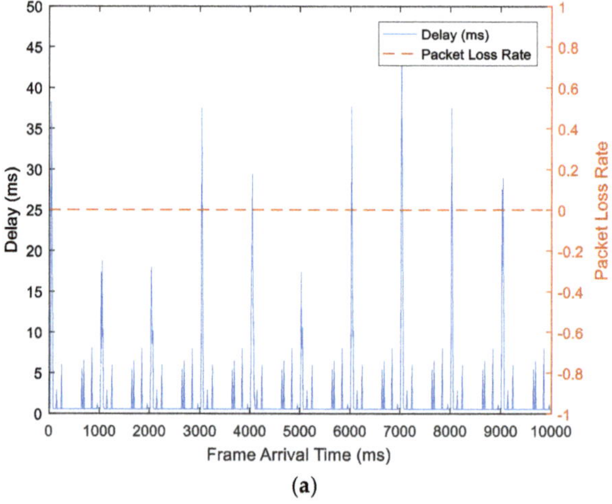

(a)

Figure 10. *Cont.*

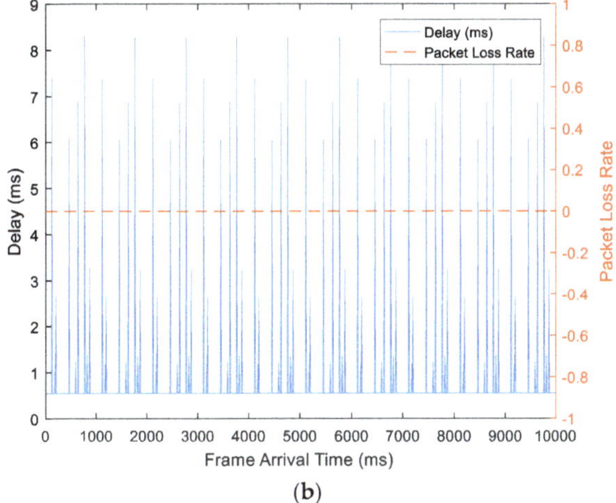

(b)

Figure 10. Delay and PLR for proposed WLAN structure with proposed schemes and delivery rate control algorithm in a severe-interference environment. (**a**) Nine interference devices with max # of retx = 2; (**b**) Nine interference devices without retx.

4. Conclusions

The proposed architecture, schemes, and rate control algorithm are able to substantially improve the delay performance in a wireless multi-user VR system. We showed that the conventional EDCA of WLAN could not fully support multi-user VR systems, because its delay and PLR were not affordable for multi-user VR system. We need to consider the proposed multi-user VR system for next-generation consumer VR devices, in order to support multi-user VR services. Unless there is an OBSS, the proposed multi-user VR system achieves very low delay and PLR, which could guarantee seamless VR UX. We observed that the retransmission procedure did not provide meaningful PLR improvement in the proposed wireless multi-user VR system. Without the need for retransmissions, by setting the maximum number of retransmission to be 0, we can significantly enhance the delay performance. Furthermore, because the proposed wireless multi-user VR system does not require modifications to the standard, the proposed scheme can be easily used with commercial WLAN chipsets that are available on the market, by adding VRCL layer proposed in this paper.

Acknowledgments: This research was supported by Basic Science Research Program through the National Research Foundation of Korea (NRF) funded by the Ministry of Science, ICT & Future Planning (NRF-2017R1A2B4003987).

Author Contributions: Jinsoo Ahn, Young Yong Kim and Ronny Yongho Kim conceived the study. Jinsoo Ahn and Ronny Yongho Kim did the literature review. Jinsoo Ahn designed the model, implemented the simulation program and obtained results. Young Yong Kim and Ronny Yongho Kim did the editing and removed the grammatical mistakes. Ronny Yongho Kim in consultation of rest of the authors, re-confirmed the credibility of obtained solutions. All authors have read and approved the final manuscript.

Conflicts of Interest: The authors declare no conflicts of interest.

References

1. He, P. Virtual reality for budget smartphones. *Young Sci. J.* **2016**, *18*, 50–57.
2. Institute of Electrical and Electronics Engineers (IEEE). 2016 Index IEEE consumer electronics magazine. In *IEEE CEM*; IEEE: Piscataway, NJ, USA, 2016; Volume 5, pp. 137–147. [CrossRef]
3. Kim, J.; Tian, Y.; Mangold, S.; Molisch, A.F. Joint Scalable Coding and Routing for 60 GHz Real-Time Live HD Video Streaming Applications. *IEEE Trans. Broadcast.* **2013**, *59*, 500–512. [CrossRef]

4.   Institute of Electrical and Electronics Engineers (IEEE). *Wireless LAN Medium Access Control (MAC) and Physical Layer (PHY) Specifications Amendment 3: Enhancements for Very High Throughput in the 60 GHz Band*; IEEE: Piscataway, NJ, USA, 2012.

5.   Institute of Electrical and Electronics Engineers (IEEE). *P802.11ay/D0.3—Wireless LAN Medium Access Control (MAC) and Physical Layer (PHY) Specifications—Amendment 7: Enhanced Throughput for Operation in License-Exempt Bands above 45 GHz*; IEEE: Piscataway, NJ, USA, 2017.

6.   Billinghurst, M.; Clark, A.; Lee, G. A Survey of Augmented Reality. *Found. Trends Hum. Comput. Interact.* **2015**, *8*, 73–272. [CrossRef]

7.   Thomas, J.; Bashyal, R.; Goldstein, S.; Suma, E. MuVR: A multi-user virtual reality platform. In Proceedings of the 2014 IEEE Virtual Reality (VR), Minneapolis, MN, USA, 29 March–2 April 2014; pp. 115–116. [CrossRef]

8.   Chang, W. Virtual Reality System. U.S. Patent 14,951,037, 11 November 2015.

9.   Institute of Electrical and Electronics Engineers (IEEE). *Wireless LAN Medium Access Control (MAC) and Physical Layer (PHY) Specifications*; IEEE: Piscataway, NJ, USA, 2016.

10.   Chang, Z.; Alanen, O.; Huovinen, T.; Nihtila, T.; Ong, E.H.; Kneckt, J.; Ristaniemi, T. Performance analysis of IEEE 802.11ac DCF with hidden nodes. In Proceedings of the 2012 IEEE 75th Vehicular Technology Conference (VTC Spring), Yokohama, Japan, 6–9 May 2012; pp. 1–5.

11.   Dikbas, S.; Altunbasak, Y. Novel True-Motion Estimation Algorithm and Its Application to Motion-Compensated Temporal Frame Interpolation. *IEEE Trans. Image Process.* **2013**, *22*, 2931–2945. [CrossRef] [PubMed]

12.   Veselov, A.; Gilmutdinov, M. Iterative hierarchical true motion estimation for temporal frame interpolation. In Proceedings of the 2014 IEEE 16th International Workshop on MMSP, Jakarta, Indonesia, 22–24 September 2014; pp. 1–6.

13.   Jeong, S.G.; Lee, C.; Kim, C.S. Motion-Compensated Frame Interpolation Based on Multihypothesis Motion Estimation and Texture Optimization. *IEEE Trans. Image Process.* **2013**, *22*, 4497–4509. [CrossRef] [PubMed]

14.   Kim, D.; Lim, H.; Park, H. Iterative True Motion Estimation for Motion-Compensated Frame Interpolation. *IEEE Trans. Circuits Syst. Video Technol.* **2013**, *23*, 445–454. [CrossRef]

15.   Tang, C.; Wang, R.; Wang, W.; Gao, W. A new frame interpolation method with pixel-level motion vector field. In Proceedings of the 2014 IEEE Visual Communications and Image Processing Conference, Valletta, Malta, 7–10 December 2014; pp. 350–353.

16.   Choi, B.D.; Han, J.W.; Kim, C.S.; Ko, S.J. Motion-Compensated Frame Interpolation Using Bilateral Motion Estimation and Adaptive Overlapped Block Motion Compensation. *IEEE Trans. Circuits Syst. Video Technol.* **2007**, *17*, 407–416. [CrossRef]

17.   Ellsworth, J.J.; Johnson, K.W.; Clements, K. Head Mounted Display Performing Post Render Processing. U.S. Patent 15,043,133, 12 February 2016.

18.   Ahn, J.; Kim, Y.; Kim, R.Y. Delay oriented VR mode WLAN for efficient wireless multi-user virtual reality device. In Proceedings of the 2017 IEEE International Conference on Consumer Electronics (ICCE), Las Vegas, NV, USA, 8–10 January 2017; pp. 122–123.

19.   Institute of Electrical and Electronics Engineers (IEEE). *P802.11ax/D2.0—Wireless LAN Medium Access Control (MAC) and Physical Layer (PHY) Specifications—Amendment 6: Enhancements for High Efficiency WLAN*; IEEE: Piscataway, NJ, USA, 2017.

20.   Welch, G.; Foxlin, E. Motion tracking: No silver bullet, but a respectable arsenal. *IEEE Comput. Graph. Appl.* **2002**, *22*, 24–38. [CrossRef]

21.   Bao, Y.; Wu, H.; Ramli, A.A.; Wang, B.; Liu, X. Viewing 360 degree videos: Motion prediction and bandwidth optimization. In Proceedings of the 2016 IEEE 24th International Conference on Network Protocols (ICNP), Singapore, 8–11 November 2016; pp. 1–2.

22.   Khorov, E.; Loginov, V.; Lyakhov, A. Several EDCA parameter sets for improving channel access in IEEE 802.11ax networks. In Proceedings of the 2016 International Symposium on Wireless Communication Systems (ISWCS), Poznan, Poland, 20–23 September 2016; pp. 419–423.

MDPI

St. Alban-Anlage 66

4052 Basel

Switzerland

Tel. +41 61 683 77 34

Fax +41 61 302 89 18

www.mdpi.com

Applied Sciences Editorial Office

E-mail: applsci@mdpi.com

www.mdpi.com/journal/applsci